Student Review Guide

Joseph J. Luczkovich, Ph. D.
and
David B. Knowles
Department of Biology

John Wiley & Sons, Inc.

Cover Photo: "Leadqueen Meadow, Purcell Range"©Larry Carver

To order books or for customer service call 1-800-CALL-WILEY (225-5945).

ISBN 0-471-21885-5

Printed in the United States of America

10 9 8 7 6 5 4 3 2 1

Printed and bound by Bradford & Bigelow, Inc.

Preface

The environment and environmental issues can be some of the most interesting subjects to study and explore. This *Student Review Guide* will help open the students' and instructors' worldview to the environmental conditions and environmental issues before us. This review guide will allow you to access the most current information on such as topics as: human population growth, deforestation, natural resources extraction, biodiversity conservation, changing land use patterns, energy resources, air and water pollution, global climate change, and environmental ethics.

The chapters follow Botkin and Keller's *Environmental Science: Earth as a Living Planet*, Fourth Edition. The sections in this review guide include "*The Big Picture*" which is an abstract or summary of the chapter concepts, followed by a section entitled "*Frequently Asked Questions*", in which we have highlighted key points from each chapter in a question/answer format. Also included in the study guide are the sections entitled: "*Ecology In Your Backyard*" which attempts to bring ecological concepts or environmental issues into the students' everyday experience; "*Links In The Library*" which is a short bibliography of pertinent publications you may be able to find in your library; and "*Ecotest*" which is a quick quiz of chapter concepts. This review guide does not replace your textbook, but will reinforce key concepts. Use it in conjunction with lecture notes, textbook readings, and other educational resources to learn about the ever-changing planet Earth. Also, be sure to spend time away from your computer and texts and simply observe and explore Nature. Relate what you see to the new information you are gaining from your environmental science course. Please let us know your opinions about this review guide by emailing us directly:

Joseph J. Luczkovich, Ph.D.: luczkovichj@mail.ecu.edu
David B. Knowles: knowlesd@mail.ecu.edu

Acknowledgments

We thank John Wiley & Sons for providing us with grants during the preparation and revision of this review guide. We wish to acknowledge and thank our editor, Cyndy Taylor, especially for her patience while we were attempting to make our deadlines. We thank Joan Petrokofsky and the staff of John Wiley & Sons, who have been very helpful in planning and soliciting feedback from users of this review guide. In addition, we wish to thank East Carolina University, the ECU Department of Biology, and the Institute for Coastal and Marine Resources at ECU for computer support, Internet connections, and printer support. Lastly, we wish to acknowledge the authors of *Environmental Science: Earth as a Living Planet*, Fourth Edition, Daniel Botkin and Edward Keller, for providing an excellent textbook to use as a foundation for writing this review guide.

Table of Contents

Chapter 1- BASIC ISSUES IN ENVIRONMENTAL SCIENCES

The Big Picture

There are some recurring themes in the book you are about to use to learn about the environment. **Human population growth** is in the exponential phase, and much of the change in the environment is driven by this massive growth. The world is becoming more **urbanized**, so problems must be solved in urban settings, not always in the wilderness. **Sustainability** (using natural resources at a rate that allows sufficient time for replenishment) is an unrealized goal, but a worthy one. The environmental problems are **global** in scale and this requires worldwide cooperation. Human **values and knowledge** play important roles in the environmental conflicts; **knowledge** or scientific data can only suggest what the alternative solutions are, but which alternative we choose is still a matter of **values** or cultural preference.

Frequently Asked Questions

What is sustainability?

- Sustainability is the ability of an ecosystem, population, or other system to perpetuate its functions or production indefinitely even with removal of environmental resources or harvests by humans.
- We can differentiate four types of sustainability:
    - 1) sustainable resource harvest;
    - 2) sustainable ecosystem function;
    - 3) sustainable economy;
    - 4) sustainable development.
- Although rapid changes are occurring, most American businesses plan with economic sustainability in mind (that is, they turn a profit or go out of business), but often do not consider the other three types of sustainable practices.

How rapidly is the human population growing?

- We are in the exponential growth phase (the most rapidly increasing phase of any population) of the human population on Earth.
- In the past 35 years, the human population has doubled from 2.5 billion to over 6.1 billion people, and it will reach 10 to 12.5 billion in the next 50 years. (See Chapter 5)
- Figure 1.3 shows the changes in population growth projected for various regions of the world. Most of the growth will occur in Asia and Africa. Europe will actually drop in population in the future, if current trends continue.

What is the carrying capacity of an environment?

- The carrying capacity refers to the maximum number of individuals of a species that can be sustained by an environment without decreasing the capacity of the environment to sustain that same number of individuals in the future.

What is the carrying capacity of the Earth for humans?

- This number is not known for certain and can only be estimated for humans. Some conservative estimates suggest that we have already surpassed the carrying capacity, and that in the future we shall experience declines in human numbers.
- The quality of life that people can expect will almost certainly decline with increased population size. It is clear that there will be some limit to growth of the human population, but the exact population size at which the Earth's carrying capacity will be reached is not known for certain.

Are there any indications that we are approaching the carrying capacity for humans?

- Yes. The recent famines in Somalia and other African nations are the best indicator that in some regions of the world, the human population has reached the environment's carrying capacity.
- Population growth in sub-Saharan Africa has begun to exceed food production (i.e., per capita food production is declining).
- Interestingly, the human population is still rising in nations affected by the famines, largely because of the reproductive inertia of human population growth, the population's willingness to live in such poor environmental conditions, and the food and humanitarian assistance provided by governments in neighboring regions (such as the U.S. and European countries).
- It is nearly certain that there will be continued declines in per capita food production in these nations. Political unrest, starvation, and armed conflict will affect this region for some years to come, until the population starts to decline.

What is the "Gaia Hypothesis"?

- The "Gaia Hypothesis" is really a concept that the environmental conditions found on the Earth (such as the oxygen-dominated atmosphere that is produced by plant photosynthesis and is absent from other planets in the solar system) are caused by the symbiotic interactions of the organisms on Earth and are self-regulating.
- Central to this concept is the idea that all species on the planet are interacting indirectly as a "superorganism" and that ecological damage in one part of the Earth will be reflected elsewhere.
- This concept of the Earth as a superorganism has been suggested by James Lovelock and Lynn Margulis (although they were not the first to propose such a concept).
- Gaia , who was the Goddess of the Earth in Greek mythology and roughly equivalent to our idea of "Mother Earth", embodies such a superorganism concept, although such a mythical figure doesn't really exist.

- Some consider the name "Gaia" unscientific and this "hypothesis" to be untestable, but the concept of self-regulation of complex systems such as the Earth's Biosphere is certainly a scientifically plausible idea.
- Self-regulation means that conditions (such as temperature) on Earth appear to be controlled or held within bounds by organisms on Earth so that these conditions remain more or less constant over time.
- If the conditions are disturbed, they tend to return to the equilibrium due to the influence of the organisms, and hence the Earth can be self-regulating.
- Increasingly, the "Gaia Hypothesis" worldview of global interactions suggests that solutions to environmental problems are global in scale, comprised of many local ecological disruptions that accumulate globally. Thus, both local and global issues are related and solutions must be sought across all spatial scales.

What is urbanization?

- The movement of people from small towns and rural areas to large cities is called urbanization.
- Developed countries like the U.S. have 75 % of the population living in urban areas, whereas undeveloped nations have only 38 % of the population living in cities.
- The environmental impacts associated with urban areas extend far beyond their borders, because all food and water must be imported into cities.
- Importation of food and water increases the environmental "footprint" of an urban population greatly, because there are many environmental costs (such as increased air and water pollution, loss of wetlands and farmlands, and the loss of wildlife habitat) associated with the intensive production and transportation of food to cities.
- Compounding this is the tendency of city dwellers to use far more food, water, and other natural resources per capita than non-city dwellers.
- Thus, environmental disruption happens more rapidly and over a greater area in a highly urbanized country than it does in a largely rural country.

What is a megacity?

- An urban area with more than 8 million people is called a **megacity**.
- The number of megacities is growing; such cities include Tokyo, Bombay, and New York City.
- In 1950, there were only two megacities (New York and London).
- There were 23 megacities in 1995, 17 in the developing world.
- By 2015 there will be 36 megacities, 23 in Asia.
- Two of the largest megacities are New York and Los Angeles.
- In 1999, Tokyo was the world's largest city, and its population will increase to 28.9 million people by 2105.

What do we mean by "values" and "science" when discussing the environment?

- **Values** are rankings or preferences that people use to choose from the available options regarding the environment.
- **Science** provides a way of knowing what the available options are regarding the environment.

- Both values and science are required to reach decisions regarding the environment.
- As an example, individuals must choose between the desire to have many children and the desire to limit the global population. Although the choice will ultimately be made based on a person's values, one can use data from population experts and medical scientists to examine options regarding how many children to have.

What is a utilitarian justification for protecting the environment?

- The need for food, shelter, and economic goods all arise from environmental resources. People who use this sort of justification for protecting the environment value resources based on market economics. (e.g., the value of forest is simply equal to the value of the timber, paper and wood products that can be made from it and sold).

What is an ecological justification for protecting the environment?

- When an environmental resource performs a role or function in an ecosystem that is not directly beneficial to an individual human in an economic sense, it may be desirable to protect it in order to preserve that function for the ecosystem by using an ecological justification.
- An example of this would be the water purification function that wetlands are capable of performing. Ecological justification is often required to protect wetlands. The economic value to humans in the form of reduced sewage treatment costs can be calculated, but this cost savings is often not considered in environmental decision-making because an individual owner of the wetland receives none of that savings as income.

What is an aesthetic justification for protecting the environment?

- Some environmental resources produce economic income just because they are pleasant places to visit or provide a scenic vista.
- A good example is Yosemite National Park, because the dramatic scenery and wildlife attract thousands of visitors each year. The economic value produced by the park is a utilitarian justification, but the beauty of nature has always been valued by people independently of the economic income. Yosemite was not set aside as a national park because it would make money for the U.S. government and private citizens living nearby, but because it was a beautiful place to experience the wonders of nature.
- Natural resources such as Yosemite are often protected using an aesthetic justification.

What is a moral justification for protecting the environment?

- Some people value environmental resources solely because of their belief that they should always be present on Earth no matter what economic value they have.
- An example might be protection of the ivory-billed woodpecker, which is possibly extinct, although it was known to exist in Louisiana in the 1930's. Scientists
- The discipline of environmental ethics focuses on such questions.

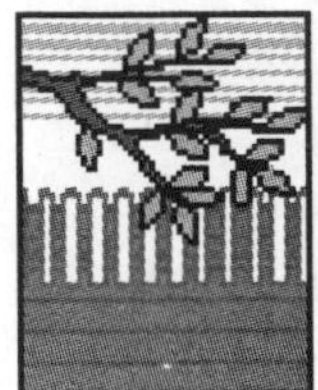

Ecology In Your Backyard

- Where in your city or on your campus do you see a choice between the old ways of doing things (non-sustainable use of resources) and the alternative sustainable approaches suggested in this chapter?
- Does your college have a recycling program?
- Do you recycle?
- Does your campus use recycled paper?
- Are you attempting to use recycled papers, reduce your paper use?
- Do you conserve fossil fuels by riding a bike or walking instead of driving?
- Does your city or campus support a bus system (buses are more sustainable than cars, because they carry more people over more miles per gallon of fuel than automobiles)?

Links In The Library

- Erlich, P. 1968. *The Population Bomb.* Ballantine Books, New York, NY.
- Lovelock, J. 1979. *Gaia: A New Look at Life on Earth.* Oxford University Press.
- Schneider, S. H. and P. J. Boston. 1991. *Scientists on Gaia.* The MIT Press, Cambridge Massachusetts. 443 pp.

Ecotest

1. Which type of sustainability is understood and practiced by most American businesses?
a. sustainable resource harvest
b. sustainable ecosystem function
c. sustainable economic principles
d. sustainable development

2. According to best estimates, how many people live on Earth at the present time?
a. 2.7 billion
b. 2.7 million
c. 5.9 million
d. 6.1 billion
e. 5.9 trillion

3. Which of these statements is true about the carrying capacity?
a. We know for certain what the carrying capacity for the human population on Earth is.
b. The carrying capacity is the maximum number of individuals that can be sustained in an area without decreasing the capacity of the environment to sustain that same amount in the future.
c. The carrying capacity for humans on Earth has been exceeded.
d. The human population is far below the carrying capacity for humans on Earth.
e. The carrying capacity for people on Earth is 6.1 billion people.

4. Which of the following might be an indication that humans have exceeded the carrying capacity of certain regions of Earth?
a. Famines in parts of Africa
b. Political unrest, revolutions, and armed conflict in the Middle East
c. The massive decrease in the birth rate and increase in the death rate in Africa
d. Population estimates that suggest a population of 11.4 billion people in 2040
e. There is no evidence that the carrying capacity has been exceeded in any region.

5. The "Gaia Hypothesis":
a. is a concept of the Earth as a self-regulating "super-organism".
b. suggests that there is a Goddess of the Earth named Gaia.
c. is a hypothesis that can be rigorously tested by scientific experimentation.
d. states that "Mother Earth" controls the biosphere's ecosystems.
e. suggests that the Earth can be damaged and it will "heal" itself.

6. What percentage of the population of all developed countries, such as the U.S., live in cities?
a. 25 %
b. 50 %
c. 75 %
d. 10 %

7. Which of the following environmental problems are associated with urbanization?
a. The loss of wetlands
b. The loss of farmlands
c. The increase in air pollution
d. The increase in water pollution
e. All of these are correct.

8. Which of the following statements that were overheard at a recent public meeting about declining fish stocks is a utilitarian justification for protecting the fish?
a. "The natural balance of the food web is changed when excessive fishing is allowed. In order to protect the entire marine ecosystem, including whales and dolphins, we should conserve fish stocks."

b. "Many people come to the coast to fish and they spend $ 1 million each year on food, drinks, ice, boats, fuel, and fishing equipment. We want to protect fish stocks so that these people will come back year after year."
c. "I have visited the public aquarium and I like to watch the fish – they are very beautiful. I think that these fish populations and their natural beauty must be protected."
d. "I am opposed philosophically to the practice of fishing. It is not humane to the fish."

9. Why do people value the environment?
a. Ecological concerns
b. Utilitarian concerns
c. Moral and ethical concerns
d. Aesthetic concerns
e. All of these are correct

10. Which of the following practices is likely to be sustainable?
a. The clear-cutting of old growth forests in the Pacific Northwest
b. The destruction of wetlands to be replaced by a sewage treatment facility
c. The construction of aquaculture ponds in mangrove ecosystems
d. The harvest of salmon from the Columbia River by a small population of Native Americans in that region using traditional methods of capture

11. The movement of people from small towns and rural areas to large cities is called:
a. sustainability.
b. urbanization.
c. migration.
d. mitigation.

12. Which of the following is *not* an example of an ecological justification for protecting the environment?
a. Saving a wetland that functions in water purification from being drained.
b. Protecting forests that provide habitat for endangered species from clearcutting.
c. A forest is protected from being cut down because it has a moral right to exist.
d. Limiting the number of fish that can be caught because they help control mosquito populations.

13. All of the following are environmental "footprints" left by a city except:
a. loss of beachfront dunes due to hotels being built in a coastal city.
b. increased levels of smog because of a new interstate.
c. higher taxes because of a need for new government offices.
d. water pollution from a sewage treatment plant.

14. Which of the following best describes the current growth phase of the Earth's human population?
a. Linear growth
b. Exponential decline
c. Sporadic growth

d. Exponential growth
e. Decaying growth

15. Human population in nations affected by famines is still on the rise. This is due to all of the following except:
a. immigration of people from surrounding areas.
b. food and humanitarian assistance of neighboring countries.
c. the reproductive inertia of human population growth.
d. the population's willingness to live in poor environmental conditions.

Chapter 2- THINKING CRITICALLY ABOUT THE ENVIRONMENT

The Big Picture

Scientific thinking involves being critical or skeptical about claims made by people. Environmentalists, government officials, corporate spokespeople, news media reporters, advertisers, politicians, and just the average citizen often make statements about an environmental issue that prove later to be unfounded, untrue, or based on incomplete or wrong information. In this chapter, the authors attempt to describe the methods used by scientists to establish knowledge and give some examples of scientific, pseudoscientific, and non-scientific thinking. In order to illustrate how the public can be misled, consider the Case of the Mysterious Crop Circles (Case Study). The media and many people were fooled into believing that the designs, which appeared overnight in agricultural fields in England, were due to various forces, including: aliens, electromagnetic fields, and whirlwinds. After much investigation, it was concluded that the designs were made by people with boards who dragged them through the fields, created the intricate patterns, and carefully covered their tracks as they went. Scientists are often some of the most skeptical people in the world, often not believing reports and explanations in the media of incredible events like the crop circles. "Tabloid" newspapers and television shows do little to improve the credibility of the news media, which often report phenomena based on incomplete data or a few non-scientific sources. Scientists are skeptical people because they rely on the scientific method to aid them in finding the truth. The scientific method uses an inductive logic tree, in which successive hypotheses are eliminated one at a time using experiments. Most people use everyday thinking to evaluate statements about the environment, which differs from the scientific method in that it is an informal method based on trial and error experiences, hearsay evidence, and common sense, not careful observation and controlled experimental data. Scientists are slow at reaching "the truth" and often state that there aren't enough data to reach a certain conclusion; what they mean is that they have not ruled out many alternative explanations for an event. In fact, "the truth" is impossible to know for certain, for science advances by disproof of hypotheses, and the disproved explanations are eliminated from being "the truth". Any remaining hypothesis or explanation is considered to "be true" until eliminated by an experimental test. The only exception to this rule is that a hypothesis or explanation, to be considered scientific, must also be falsifiable, that is the phenomenon or proposed explanation must be observable by people and the hypothesis must be phrased in a way that it is possible for an experiment to show it to be false. Because of the falsifiablilty rule, most explanations involving unobservable events (such as aliens or supernatural beings) are eliminated from consideration by scientists immediately.

Frequently Asked Questions

What is science?

- Science is an organized human activity, based on observation of nature and the application of the scientific method that has been devised as a way to assemble falsifiable statements known not to be false.
- Scientific statements have a high probability of being true, although they are not known to be true without a doubt.
- Science represents all of human knowledge and this is typically written down in scientific books and journals.

What is the difference between science and technology?

- Science is a search for understanding of the natural world, whereas technology is the control of the natural world for the benefits of humans.
- Science depends on new tools and inventions to improve the ability of scientists to make observations of nature.
- Technology often advances because of scientific discoveries that were made in pursuit of knowledge for its own sake. For example, scientists using the light microscope were limited previously to seeing things no smaller than 1/10 of a micrometer and magnifying them 1000 times. Now, with the invention of the electron microscope, they can see things that are much smaller, because the electron microscope can magnify objects 100,000 times. Because an understanding of electron beams and focusing them required scientific investigation, the technology that improved microscopy was dependent upon science.
- Thus, science and technology are interdependent human activities, and one promotes the other. They are often used interchangeably by people in discussing environmental issues; however, they are different but related areas of human activity.

What isn't science?

- Statements about the ultimate purpose of life, the existence of supernatural beings (including God), statements on human values, beauty, good and evil are not considered part of science.
- Scientists often have opinions on such topics, but that does not make their opinions scientific fact.
- Statements about supernatural beings and religious belief systems are very important to humans, they just are not part of science.
- **Pseudoscience** - Literally meaning "false" science, this approach can be very common in debates about environmental issues. People may attempt to use untested or disproved ideas and hypotheses to explain various phenomena.
- For example, 25 % of Americans believe in astrology, that the motion of the planets and stars can affect human behavior. Astrology is pseudoscience; astronomy (the study of the motion of the stars and planets) is not. The position of any planet can be

predicted with high accuracy and precision, but the behavior of people cannot be predicted by knowing these positions.

What is the scientific method?

- The scientific method consists of four steps:
    1. Observations based in nature
    2. Hypothesis formation based on the observations
    3. Experimental tests of the hypothesis
    4. Falsification or acceptance of the hypothesis
- After these steps have been completed, the hypothesis can be discarded (if false) or added to a scientific theory (if accepted). A new cycle of hypothesis formation and testing then begins.

What kind of reasoning is used in everyday life?

- Most people reason using common sense, which is more than adequate for daily decisions of what to buy, what to wear, what to do in a given situation.
- These decisions are guided by previous experiences and an expectation that a future event will be similar to what occurred in the past.
- However, this everyday knowledge and reasoning is tolerant of imprecision, is acquired by trial and error methods, is not validated by formal methods of testing, is often taken on faith or assumed based on what others say, (especially parents or experts) and is based on subjective values or beliefs.
- Most of the time, these limitations do not significantly influence the outcome of the decision, or people don't care if they do.

What are the assumptions that all scientists make?

- Events in nature follow patterns.
- There are universal patterns or rules.
- General conclusions can be drawn from specific events; this type of reasoning is called inductive logic.
- These general conclusions can be tested; tests can falsify or disprove the generalizations (or hypotheses).

What is scientific "proof" and how does it differ from "proof" in mathematics and in everyday life?

- Science cannot provide definitive proof; scientists can only make statements that have a high probability of being true.
- Science advances by disproving false hypotheses: any hypotheses left after testing are considered to be true for the time being.
- A proof in mathematics involves using deductive logic to show that given initial starting premises, a certain condition follows. This kind of proof is definitive, but it is not the same as scientific proof.
- When someone says that something has been proven scientifically, it is better to think of this as a statement that is likely to be true 95 - 99% of the time.

What is a hypothesis?

- A hypothesis is a statement (not a question) that might explain some event or natural phenomenon.
- Usually, hypotheses propose a cause for the event, in a conditional sense: "If it is warm, clams will grow faster than if it is cold."

What is a theory?

- A theory is a collection of unfalsified hypotheses.
- A theory thus is a collection of statements with a high probability of being true, which are put together as a narrative (or a story).
- Theories are never proven in science, but they can be disproved if certain key hypotheses that are central to the theory can be disproved.

What is an experiment?

- An experiment is a controlled comparison study among experimental groups in which all factors that could influence the outcome variables are the same among groups being compared except for one or two factors that are intentionally varied.
- For example, in an experiment to test the hypothesis that growth rate of a certain species of clam is faster at high water temperatures than at low temperatures, an aquaculturist could intentionally vary water temperature and the growth rate of clams could be monitored among groups of clams exposed to different temperatures.
- For this to be a controlled experiment, all other factors that could affect growth (food availability, sunlight, dissolved oxygen in the water, salt content or salinity of the water, substrate type, etc.) must be the same across the groups of clams; only temperature can be different among groups.
- Experiments may be done in both the laboratory and the field, but normally it is easier to control temperature and salinity in the laboratory.
- Field experiments are often more variable, but are required to demonstrate that the relationships among variables discovered in laboratory studies are valid in the natural world.

What is the difference between inductive and deductive reasoning?

- **Inductive** reasoning is used when one makes many observations and draws a general conclusion from them, usually a statement in the form of a hypothesis to be tested in an experiment. This is the kind of logic that scientists most often use to make observations and formulate hypotheses.
- **Deductive** reasoning is a form of logic in which general premises or statements are taken to be true and specific conclusions can be made based on the premises. Deductive logic does not require that the premises are true, but rather that the logic is sound in reaching a conclusion.
- An example of each type:

  1) **Inductive:** Many observations of swan color are made, and so far all swans observed are white. By the inductive approach, we conclude that all swans are

white (even though a new observation of a black swan could disprove this general conclusion).
2) **Deductive:** Premise 1: All swans are white; Premise 2: We observe a black swan-shaped bird; therefore, we conclude that this new bird is not a swan.

Aren't all scientific measurements made without error?

- No. All scientists realize that every measurement is an approximation.
- All measurements in science need to have an error or uncertainty value associated with the measurement.
- Some measurements are more precise than others, but they are all estimates of the true value.
- One source of potential error is called measurement error, in which the scientist's ability to measure is limited by the tool used in measuring, the scientist's eye, or other senses used to make the measurement.
- The true value will always be unknown, because of measurement error and other random errors. The true value is approached as the sample size (or number of measurements) increases. As sample size increases, the associated random error gets smaller.
- Example: The temperature at which the Space Shuttle O-rings fail is -1 °C (30 °F) $\pm$1 °C or colder. If the temperature at launch time is 0 °C (32 °F), then the shuttle should not be launched. Even though 0 °C is above -1 °C, the temperature is within the error range in which it will fail, according to the engineers that built the O-rings.

What is the difference between accuracy and precision?

- **Accuracy** is the degree to which a measurement approaches the true or known value. This can be assessed for measurements in which we have a standard true value that is agreed upon by all scientists. For example, seawater has an agreed upon standard value of 35.0000 parts per thousand salinity or salt content. At sea level, the boiling point of water is 100.000 °C (212.000 °F). If you measure the temperature of water boiling and your thermometer reads 99.000 °C (210.200 °F), then your thermometer is inaccurate.
- **Precision** is the degree of exactness with which a measurement is made. If your thermometer is marked in 0.1 °C (0.2 °F) increments, then it is less precise than another thermometer which reads in 0.001 °C (0.002 °F) increments.

What is an experimental error?

- This is the error associated with running an experiment (see What is an experiment?).
- The experimental error is normally estimated by measuring the disagreement within a group treated the same way in an experimental treatment (statistically, this is done by calculating its variance).
- Thus, in the example above, all clams in a temperature treatment group should grow at exactly the same rate, but they rarely do in actual experiments. The reason is that there is error in measuring the growth rate (measurement error associated with measuring the size or weight of the clams) and there is random or unassigned error

(the temperature varies ever so slightly for each clam, the clams are all unique individuals with their own metabolic rates, etc.).

- The experimental error, which is the combined measurement and random errors, can be seen in the fact that all the clams will be measured growing at a slightly different rate within a treatment, but the variation within the treatment group will be less than the differences in growth rate between groups of clams grown at different temperatures. So, experimental error gives the variation among individual clam responses as well as the error associated with the temperature effect.

What is the difference between an independent and a dependent variable?

- A **dependent** variable is the response variable or the measurement that will most likely respond to an experimental treatment. In our clam growth experiment, this would be the growth rate measured as g of body mass (excluding the shell) added/day.
- The **independent** variable is the variable in an experiment that is intentionally varied. In the clam example, the independent variable is temperature.

What is meant by an operational definition of a variable?

- When scientists are describing their methods for conducting an experiment, they must produce a list of the variables that were measured and a definition for how the measured variables were obtained. These are the **operational definitions.**
- In our clam example, the way that growth was measured (g of shell-less body mass added/day) must be specified: the clams were grown at each of the temperatures for 14 days, and the biomass increase for each treatment group was measured by weighing the shell-less biomass of the clam, then subtracting the average shell-less biomass of a random sample of 10 clams in each group taken at the start of the study, and finally dividing by 14. This is an example of a **quantitative variable**.
- A **qualitative response variable** would be to use categories such as small, medium, or large to group the clams, and count the number of clams that fell into each group for each temperature treatment.

What is probability?

- Probability is the number of successful trials divided by the number of attempts. Success is defined as the outcome of interest.
- For example, we can estimate that 50 % of the clams in our experiment will die if the temperature gets close to 30 °C. (This is called the thermal maximum, and now that we know of its existence, we would be forced to reevaluate or reword our hypothesis stated above. How might it be changed?) We could define successful outcomes as those in which the clams survive.
- Thus, the probability of a clam surviving would increase as we decreased the temperature below 30 °C, but the probability would decrease as we approached 30 °C. If 30 °C is exceeded, all the clams die (0.00 probability of survival):

Temperature °C	No. Clams Tested (No. of Attempts)	No. Clams Surviving (No. of Successes)	Probability of Clams Surviving at Temp.
25	100	99	0.99
30	100	50	0.50
35	100	0	0.00

Note: In these types of bioassay experiments, the LT-50 (Lethal Temperature 50) thermal maximum is reported, that is the temperature at which 50 % of the population dies.

What is a conceptual or mathematical model?

- A model is a "deliberately simplified construct of nature" that allows a scientist to incorporate unfalsified hypotheses into a larger framework of ideas.
- Some models are conceptual models, represented in a diagram form (see any of the diagrams in Chapter 4 on biogeochemical cycles).
- Others are quantitative models or mathematical models, which use some of the estimated rates from experiments and measurements taken in nature to predict the future.
- For example, if global warming causes the water temperature where clams are living to increase, our experimental data can be used to predict the growth rate of the clams at that warmer temperature. They may grow faster (but so may their predators, and as we have seen, there will be a temperature which causes a decline in growth or death of the clams).
- Computers can be programmed to calculate all of the quantitative data we have gathered on many different species into a large mathematical model, and we simultaneously calculate the responses of all the organisms in an ecosystem.
- These types of models are becoming more commonplace in the scientific prediction of the consequences of global warming and other global change phenomenon.

Ecology In Your Backyard

- Find an article in your local newspaper or a magazine that discusses a scientific study of an environmental issue.
- What hypothesis is being tested by the scientists?
- Did the scientists eliminate or disprove any of the alternative explanations for the discovery?
- Can you think of any alternative explanations?
- If no hypothesis is stated in the article, write a hypothesis and a series of alternative hypotheses down. How could you test each alternative?

Links In The Library

- Gleick, J. 1987. *Chaos: Making of a New Science*. Penguin Books, New York. 352 pp. A book about the developments in Chaos theory, an emerging field of study in physics, chemistry, biology, and earth sciences. This book explains why computer models can never be perfect predictors of what will happen in nature.
- Feynman, R., R. B. Leighton, M. Sands. 1963. *The Feynman Lectures on Physics*. Volume I. Addison-Wesley, Reading, MA. See Pages 2-1 - 2-2.
    - Ideas about physics, nature, observation, description, and experiments using the scientific method.
    - Richard Feynman was a physicist who lectured extensively about science and the scientific method.
    - As an example of what scientists do, Feynman used the following scenario: imagine that you have never played chess before and you are watching a match being played (this could be any game you are familiar with). How could you figure out the rules from simply observing the players moves? When you have watched enough games and made careful notes, you probably would have a good idea of what each piece can do. At some point, someone will "castle" - a move that you could never have anticipated from previous moves of the king and the rook - and this will puzzle you. In a similar way, scientists must find patterns in nature and explain unanticipated results. If a scientist is fortunate, he or she will perform an experimental manipulation - move a chess piece - and predict the result for each piece. If it is as anticipated, great! If a castle follows - well, that's science! You need to make more observations, make more hypotheses, and conduct more experiments.
    - See also the BBC Videotape: "The Pleasure of Finding Things Out" about Feynman
- Harte, J. 1985. *Consider A Spherical Cow*. William Kaufmann, Inc, Los Altos, California.

Ecotest

1. What caused the mysterious crop circles in corn fields in England?
a. It has not been determined.
b. Science cannot prove anything for certain, so it will always be unknown.
c. Alien spaceships landing at night made them.
d. Magnetic forces made them.
e. Humans made them.

2. Scientific statements must:
a. always be true.
b. be falsifiable.
c. be based on experiments alone.
d. be regarded as false until proven true.
e. All of these are correct.

3. Scientific statements always have a high ___________________.
a. degree of certainty
b. degree of accuracy
c. probability of being true
d. degree of precision

4. ____________________ is the control of the natural world for the benefit of humans.
a. Science
b. Technology
c. Environmental science
d. The Scientific Method

5. Based on the data from a laboratory clam growth experiment given in the table below, which of the following is a non-scientific statement?

Temperature treatment	Clam Growth (g added/day)	Experimental Error (g/ day)
10 °C	0.0	$\pm$ 0.1
15 °C	0.5	$\pm$ 0.3
20 °C	1.0	$\pm$ 0.2
25 °C	1.5	$\pm$ 0.2

Note: The weights of the clams were measured on a balance with a $\pm$ 0.1 precision. All water salinities were maintained at 35 parts per thousand.
a. The clams grew fastest at 25 °C.
b. The increased water temperature caused the clams to grow fastest at 25 °C.
c. The cause of increased clam growth is in all probability due to increased water temperature.
d. The increased growth of clams at higher temperatures occurred because warm water makes the clams happiest.

6. Based on the above clam growth data, at what temperature would you expect a clam farmer to grow clams to market size the fastest?
a. 10 °C
b. 15 °C
c. 20 °C
d. 25 °C

7. At which temperature is the experimental error the smallest?
a. 10 °C

b. 15 °C
c. 20 °C
d. 25 °C

8. In the above experiment, the ________________ is ± 0.1 g/day for clam growth rate.
a. accuracy
b. precision
c. probability
d. experimental error

9. Which of these is the independent variable in the clam growth experiment described above?
a. Temperature (°C)
b. Salinity (parts per thousand)
c. Clam growth rate (g added/day)
d. All of these are independent variables.
e. None of these are independent variables.

10. Calculate the probability of a clam surviving when the water temperature is 29 °C, if you were given the following data: 50 clams were held at 29 °C and 28 of them survived.
a. 0.28
b. 0.56
c. 0.44
d. 0.73

11. The definition, "a deliberately simplified construct of nature", best identifies which of the following concepts?
a) Hypothesis
b) Algorithm
c) Model
d) Operation
e) Variable

12. Which of the following is an example of a qualitative response variable?
a) The growth of fish (g/day) that are fed a certain diet over the span of a month.
b) The number of dead fish in a fish kill.
c) Separating apples based on whether they are ripe or not.
d) The amount of pesticides in a water supply, measured in parts per million.

13. Which statement about theories is false?
a) A theory is a collection of unfalsified hypotheses.
b) A theory is a collection of statements with a high probability of being true.
c) A theory never changes once it has been stated.
d) Theories are never proven in science.

14. Which of the following is an example of inductive logic?
a) All fish that have been examined have scales, thus if an organism has scales, it must be a fish.
b) All fish that have been examined have scales, thus fish always have scales.
c) All snakes that have been observed have scales, thus if an organism has scales, it must be a snake.

15. Which of the following is a good example of the use of pseudoscience?
a) The moon is full tonight, thus we should expect a high tide now.
b) Jupiter is aligned with Mars, thus I will win on the roulette wheel tonight.
c) The sun is highest in the sky at noon, thus I might get a sunburn if I stay out in the sun then without protection from ultraviolet rays at that time.

Chapter 3 - SYSTEMS AND CHANGE

The Big Picture

A **system** is any interconnected set of components that acts as a whole. In **ecosystems,** the components are organisms (plants, animals, and microorganisms). Organisms are interconnected by the wastes that they produce, the chemicals that they assimilate from the environment, and by the food webs (feeding interconnections) of which they are a part. Ecosystems are naturally changing biological systems that approximate a **steady state** over a long period of time. A **steady state** condition means that there is no net change in the system: input = output. This idea, that the system is in an **equilibrium state**, is popularly expressed as the "balance of nature". Although equilibrium conditions do not always occur in ecosystems at any given time, they appear to be in a steady state over a long period of time. When one component in an ecosystem is pushed from the steady state conditions by human or natural **disturbance**, the other parts change as well, acting together as a system. Examples of natural disturbances include floods, hurricanes, storms, and fires. Natural changes in ecosystems are due to random disturbances; these are in turn controlled by **positive** and **negative feedback loops.** An example of natural changes being mistaken for changes caused by humans can be observed in Amboseli National Park in Kenya (see the Case Study). When forests disappeared, Masai herdsmen were accused of causing the environmental damage. After careful study, it was determined by scientists that the forests disappeared because rainfall increased (due to a long-term change in climate), bringing salty groundwater to the surface. Thus, even though it was widely believed that the Masai were encouraging overgrazing of the land, humans were not to blame for the ecosystems change. Such changes in ecosystems are often subtle and difficult for humans to observe, because they occur over many years. Only careful scientific measurements over a long period of time and experiments can differentiate human impacts from natural changes in ecosystems.

Frequently Asked Questions

What is a system?

- A system is any interconnected set of components that acts as a whole.
- Systems may be open (no boundaries, material moves in and out) or closed (bounded, no input or output).
- Ecosystems are typically open systems, but are studied as having artificial boundaries.
- The Earth (Biosphere) is a closed system, for all practical purposes.

What is a positive feedback loop?

- These are feedback loops in which an increase in output leads to an increase in input (see Figure 3.6)

What is a negative feedback loop?

- These are feedback loops in which an increase in output leads to a decrease in input (see Figure 3.8).

What is exponential growth?

- Exponential growth is a constant rate of growth in which the number of organisms at any time is a multiple of the number present in the previous time period. It is an example of positive feedback.

What is doubling time?

- Doubling time is the time necessary for the quantity of whatever is being measured to double. For populations, this is normally measured in years.

What is the Concept of Environmental Unity?

- The concept of environmental unity suggests that "everything is hitched to everything else" in an ecosystem.

What is Uniformitarianism?

- The concept of uniformitarianism suggests that the history of the Earth can be found in fossils trapped in the layers of rocks, with the lowest layers containing the oldest rocks and fossils.

What is steady state?

- When there is no net change in the size of a system (input = output), then the system is said to be in a steady state or at equilibrium.
- When freshmen entering the university are equal to the seniors leaving the university, enrollment is in a steady state.

What is residence time?

- Residence time is the average amount of time it takes for materials in a pool of resources to cycle through the system in a steady state. It is computed as the total system size divided by the rate of flow of materials through the system.
- If the university has 10,000 students, and each 2,500 students enter as freshmen and 2,500 graduate as seniors, it is in a steady state with a inflow and outflow of 2,500 students. The average residence time is 10,000 students (total enrollment) divided by 2,500 students/year (flow of students through the system) or 4 years.

What is the "balance of nature"?

- In this popular concept, ecosystems tend toward a steady state or equilibrium condition when humans are absent. Species are replaced as they die off, populations are unchanging.
- Humans are often viewed as upsetting the "balance of nature".

Is the equilibrium concept an appropriate one for ecosystems?
- If one observes an ecosystem for a long-period of time, it becomes clear that steady state conditions may only prevail for short periods of time.
- Natural changes can occur in the inputs to and outputs from the system, as well as human-induced changes to these inputs and outputs. For example, rainfall and climate may change over time, which can move the system out of steady state.

What is an ecological disturbance?
- Ecosystems are being displaced from a steady state on a regular basis by **ecological disturbance**. Examples of natural disturbance includes hurricanes, floods, and wildfires. Humans can cause disturbances as well (pollution, habitat destruction).
- After disturbance, ecosystems can change from one steady state to another, or return to the original steady state. This variability in steady state is the ecological concept of multiple stable states for ecosystems.
- Disturbance has been shown to be a requirement for some ecosystems to maintain species diversity.

Don't natural disturbances prevent steady state conditions from ever being reached?
- Some ecologists think that ecosystems never reach an equilibrium because of frequent disturbances. For example a high surf zone (like on North Carolina's Outer Banks beaches) is a habitat in which disturbance is frequent. In such systems, equilibrium may never be reached.

How old is life on Earth?
- The Earth formed approximately 4.6 billion years ago, when a solar nebula collapsed, forming protostars and planetary systems.
- Life began 3 billion years ago, when the Earth was 1.6 billion years old.
- Fossils are an indication that life existed 3 billion years ago
- Life on Earth (biota) profoundly changed the environment, adding oxygen to the atmosphere, for example.

What are extinction rates and how have they changed with human population growth?
- Extinction rates measure the number of species (kinds of organisms) that have disappeared from Earth over time.
- The rates of extinction of mammals and birds have increased as humans have become more abundant on Earth, suggesting that humans cause the extinction of some species.

What does biota mean?
- The biota are the living things on Earth (plants, animals, and microorganisms).

What is the Biosphere?

- The Biosphere is the surface layer of the Earth in which the biota can be found, including parts of the lithosphere (rocks), the hydrosphere (water) and atmosphere (air).

What is an ecosystem?

- An ecosystem is a community of organisms and its local nonliving environment in which matter (chemicals elements) cycles and energy flows.
- Ecosystems may be large or small, natural or artificial, and have their boundaries variously defined, but they all share the same feature: biota that are the cause of energy flow and material cycling.

What is the Gaia hypothesis?

- The Gaia hypothesis is the superorganism concept for Earth ecosystems, developed by James Lovelock, and named for the Greek Goddess of Earth, Gaia. This hypothesis suggests that there is a giant positive feedback loop in which some lifeforms (like plants that produce oxygen) promote the conditions for other lifeforms (like animals that consume it). (see *Frequently Asked Questions* in Chapter 1)

Is the Gaia hypothesis a testable scientific hypothesis?

- No, because it is a worldview rather than a hypothesis. This worldview is comprised of individual hypotheses about positive feedback loops that are testable hypotheses.

Ecology In Your Backyard

- Have you been aware of ecosystem changes in your town or region?
- What are the suggested causes of the ecosystem changes?
- Are humans blamed for the change?
- Could the changes be due to natural, long-term changes in climate?
- Are ecological disturbances to blame for the changes?

Links In The Library

- Jones, P. and T. Wigley. 1990. *Global Warming Trends*. Scientific American. August 1990, pages 84-91.
- Repetto, R. Deforestation in the tropics. *Scientific American*. April, 1990, pages 36-42.
- *Consequences: The Nature and Implications of Environmental Change*. - This excellent journal contains many articles on how humans are transforming the Earth. It has only been in publication for a few years, but there about sixteen articles, all of which relate to

the issues in this chapter and in later chapters. Your library may not get this series, but it is distributed **free of charge**, so tell your librarian. These articles are also available online on the internet. To obtain any of these articles, write or fax the editorial office of *Consequences*:

John A. Eddy, Editor, Saginaw Valley State University, 7400 Bay Road, University Center MI 48710 *jeddy@tardis.svsu.edu*, fax: (906) 387-2932

Contents of Volume 1, Number 1, 1995

- Trends in U.S. Climate during the Twentieth Century
- America's Water Supply: Status and Prospects for the Future
- Past and Present Land Use and Land Cover in the U.S..

Contents of Volume 1, Number 2, 1995

- Global and U.S. National Population Trends
- Impacts of a Projected Depletion of the Ozone Layer
- Potential Impacts of Climate Change on Agriculture

Contents of Volume 1, Number 3, 1995

- The Environment Since 1970
- Climate Models: How Reliable are their Predictions?

Contents of Volume 2, Number 1, 1996

- Greenhouse lessons from the Geologic Record
- How Bountiful are Ocean Fisheries?
- The Sun and Climate

Contents of Volume 2, Number 2, 1996

- Ending Hunger: Current Status and Future Prospects
- Impacts of Introduced Species in the United States
- Population Policy: Consensus and Challenges

Contents of Volume 3, Number 1, 1997

- The Case of the Missing Songbirds
- Do We Still Need Nature?

Contents of Volume 3, Number 2, 1997

- Climate, Ecology, and Human Health
- Keeping Watch on the Earth: An Integrated Global Observing Strategy

Contents of Volume 4, Number 1, 1998

- From A Carbon Economy To A Mixed Economy: A Global Opportunity
- The Carbon Cycle, Climate, and the Long-Term Effects of Fossil Fuel Burning

Contents of Volume 5, Number 1, 1999

- The Extreme Weather Events of 1997 and 1998
- Beyond Kyoto: Toward a Technology Greenhouse Strategy

Contents of Volume 5, Number 2, 1999

- El Niño and the Science of Climate Prediction
- The Great El Niño of 1997 and 1998: Impacts on Precipitation and
- Temperature
- The Application of Climate Information

Ecotest

1. A (An) ______________ is any interconnected set of components that acts as a whole.
a. ecosystem
b. system
c. steady state
d. food web

2. A (An) ____________________ feedback loop is one in which an increase in output leads to a decrease in input.
a. positive
b. negative
c. random
d. equilibrium

3. ___________________ prevents equilibrium conditions from ever being reached.
a. Ecological disturbance
b. Exponential growth
c. Steady state conditions
d. Uniformatarianism

4. The concept of Environmental Unity suggests that:
a. everything is hooked to everything else in an ecosystem.
b. the steady state conditions always are met.
c. there are layers and layers of fossil rocks that tell the history of the Earth.
d. the Earth functions as a superorganism.

5. What is the Gaia hypothesis?
a. Everything is hooked to everything else in an ecosystem.
b. The steady state conditions always are met.
c. There are layers and layers of fossil rocks that tell the history of the Earth
d. The Earth functions as a superorganism

6. The ___________________ is a layer of the Earth in which all the biota live.
a. hydrosphere
b. lithosphere
c. atmosphere
d. biosphere

7. Why should humans care about extinction rates?
a. The Earth's age can be determined from them.
b. Because the rates have declined recently, suggesting that over population by species may occur.
c. Because the rates have recently increased, suggesting that humans may be the cause of increased extinctions.
d. Because they are a good indicator of the impacts of global warming.

8. The amount of time that it takes materials to cycle through a system is called ________________________.
a. doubling time
b. residence time
c. geological time
d. cycling time

9. Which of the following is most likely to grow at an exponential rate?
a. The human population
b. A savings account
c. An unpaid credit card account
d. All of these are correct.

10. What caused the decline of forests in Amboseli National Park in Kenya?
a. A long-term increase in rainfall
b. The Masai herdsmen and their cattle
c. Elephants migrating through the region
d. Timber harvesting by large corporations

11. When conditions in a system are in a "steady-state", this means:
a. nothing at all is changing in the system.
b. the overall change in the system is zero, where input equals output.
c. the system is in constant flux, with both input and output changing in relation to one another.

12. An example of a negative feedback loop is:
a. a bank account where the more money you put in, the more you will earn.
b. a situation where a bicycle accident leads to a car swerving off of the road, which runs into a building and catches it on fire.
c. the more often you cut the grass, the faster it grows, so you have to cut it more often in the future.

d. the harvest of fish is large when the population is large, then drops over time as fish fail to replenish themselves naturally at the rate at which they are removed.

13. Which of the following statements about the Gaia Hypothesis is false?
a. It is not a testable hypothesis; but rather an opinion held by both scientists and non-scientists.
b. It is not a hypothesis, but rather a worldview, comprised of several individually testable hypotheses.
c. Hypotheses about positive feedback loops such as the Gaia Hypothesis cannot be experimentally manipulated.
d. It is a hypothesis that would be very difficult to test, because the experiments would have to be global in scale.
e. All of the above are false.

14. Calculate the residence time for a molecule of water in a lake (1,000,000 m^3 volume) that has a stream flowing in at a rate of 1000 m^3/day.
a. 1000 days
b. 1 year
c. 10 days
d. 1 day

15. Which of the following is the name given to the sphere in which all life exists on Earth?
a. hydrosphere
b. lithosphere
c. atmosphere
d. biosphere

Chapter 4- THE BIOGEOCHEMICAL CYCLES

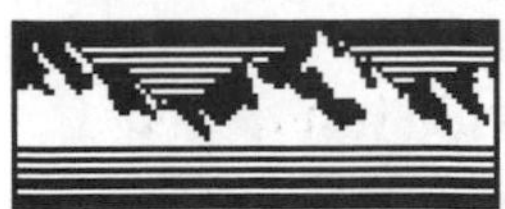

The Big Picture

Nutrients are elements or compounds that are essential for life. Nutrients, like all other elements in the Earth's crust, are finite and are recycled by natural processes. Nutrients may cycle between the lithosphere, hydrosphere, atmosphere, and biosphere. Thus, nutrient cycling is also referred to as biogeochemical cycling. Energy for biogeochemical cycling is derived in part from the solar radiation, which drives photosynthesis, and atmospheric and oceanic circulation. Energy for biogeochemical cycling is also derived from geologic processes such as subduction and uplift which are associated with plate tectonics and weathering, erosion, and rock formation, which are part of the rock cycle. In order for organisms to survive, grow, and reproduce successfully, they must be supplied with nutrients in the appropriate concentrations and at the appropriate times. Human activities have greatly altered the Earth's landscape and consequently biogeochemical cycling. The global carbon cycling has been altered by deforestation, wood burning, and the combustion of fossil fuels. Nitrogen and phosphorous cycling has been altered by agriculture, industrial processing, and urbanization. Lake Washington, near Seattle, WA, was polluted by phosphorous that was added by sewage treatment plants (see Case Study). Excessive inputs of these nutrients into waterways has resulted in eutrophication and a general decline in water quality in many waterways.

Frequently Asked Questions

What is a biogeochemical cycle?

- A biogeochemical cycle is a pathway that a chemical follows as it cycles through the biota, soil, water, and atmosphere .
- Models of biogeochemical cycles may be developed for global cycles or local ecosystem cycles (Figure 4.3).
- The cycle consists of reservoirs that are connected by pathways. In diagrams or models, the reservoirs are depicted as boxes and the pathways are depicted as arrows (Figure 4.4).

What are matter and energy?

- Matter is material that makes up our physical and biological environments; it has mass and occupies space. Matter is comprised of chemical **elements** that are made of particles called atoms, with subatomic particles: neutrons, protons, and electrons (Figure 4.2).
- Energy is the ability to do work. Energy is conserved in any system, but it can be converted from one form to another (See Chapter 16).

What elements are involved in biogeochemical cycles?

- Twenty-four of the 103 known elements are involved in biogeochemical cycles (Figure 4.5).
- Because these 24 elements are used by organisms, they are considered **nutrients**.
- Those nutrients needed in relatively large amounts by all organisms are called macronutrients
- The "Big Six" macronutrients are: Carbon (C), Hydrogen (H), Oxygen (O), Nitrogen (N), Phosphorous (P), and Sulfur (S). These are usually needed as structural components or for common biochemicals in organisms.
- Those nutrients needed in relatively small amounts or by some organisms are called micronutrients. These are usually needed as coenzymes or for very specific functions.

How can nutrients be limiting factors?

- If a required nutrient is present in concentrations so low that the metabolic needs of an organism, population, or species cannot be met, the nutrient can be considered a limiting factor.
- Conversely, if a nutrient or other element is present in a toxic concentration, growth and survival may be limited.

What are some general concepts pertaining to biogeochemical cycling?

- Biogeochemical cycling is essential for the continuation of life.
- Elements with a gas phase (e.g., oxygen and nitrogen) usually cycle most rapidly.
- Elements that readily become immobilized by geological process cycle slowly (e.g., phosphorous or carbon in fossil fuels).
- Human activities and technologies have modified the Earth's landscape and altered biogeochemical cycling rates.

What geological processes are involved in nutrient cycling?

- The geologic cycles include the **tectonic cycle**, the **hydrologic cycle**, the **rock cycle** and the **biogeochemical cycles** (Figure 4.6)
- The **tectonic cycle** redistributes nutrients and other elements within the Earth's crust (lithosphere) as a consequence of shifting tectonic plates. The slowly shifting tectonic plates that comprise the Earth's crust cause physical and chemical environmental changes (Figure 4.7). Physical changes include the alteration of global atmospheric and oceanic circulation patterns, mountain uplift and rift formation. Chemically, crustal materials are changed when they are exposed to the extreme heat and pressure associated with burial or subduction along plate margins.
- The **hydrologic cycle** is driven by the physical processes of **evaporation** and **condensation**. Water evaporates from the ocean, lake, and stream surfaces, soil, and vegetation (Figure 4.8). Atmospheric water vapor is then redistributed by **convection currents** (ascending air currents), local and global winds. As the air rises, it cools at the rate of 1°C per 100m ascent; this increases the **relative humidity.** When relative humidity reaches 100 %, **saturation** occurs, which is the point at which condensation

occurs. Water vapor then condenses around dust and other particles, which are water droplet nuclei. The aggregation of such water droplets is called a cloud. Condensation causes heat to be released (540 calories/g - 596 calories/g); this is called the **latent heat of condensation.** This release of energy causes the cloud to rise to higher elevations. When water droplets in a cloud coalesce in the chaotic environment at 30,000 to 50,000 feet, they form raindrops (1 million droplets =1 raindrop). Rain and other forms of precipitation returns the water to the oceans or land surfaces. Flowing water erodes land forms and redistributes dissolved minerals and other elements.

- The **rock cycle** is dependent upon the tectonic cycle and the hydrologic cycle. The rock cycle entails the transformation of parent rock material (formed by tectonic processes) into weathered rock material (weathered by water and other geologic, climatologic, and biologic processes) (Figure 4.9). Igneous rocks which are formed from molten magma and metamorphic rocks which have been subjected to great heat and pressure within the Earth's crust are uplifted, weathered, and their fragments are eroded. Eroded rock fragments and organic remains of plants and animals form soil and/or sedimentary rocks. Reburial of igneous, metamorphic, and sedimentary rocks by tectonic processes continues the rock cycle.
- The incorporation of these three geological cycles with the biological transformations that occur gives rise to the biogeochemical cycles, which are discussed below.

What is the role of ecosystems in biogeochemical cycling?

- An ecosystem consists of a community of organisms and their nonliving environment.
- One aspect of ecosystems is the processing of "nonliving" materials (i.e., nutrients) by the biota of an ecosystem. Most of the energy used to process nutrients in ecosystems is derived from solar radiation.
- Nutrients may be cycled internally with the same ecosystem, such as the recycling of nutrients from decomposed leaves by the trees that produced the leaves.
- Ecosystems are open systems, thus exchange materials or nutrients with other ecosystems.
- Nutrients may be translocated from one ecosystem to another via such means as flowing water, air movement, and migrating animals.
- Ecosystems that export a relatively large portion of their nutrients are called leaky ecosystems (e.g., a tidal salt marsh that exports nutrients in detritus to adjacent estuaries with each ebbing tide).

How does the carbon cycle function?

- Carbon is the basic structural element of all organic molecules, thus essential to all life.
- Even though life is carbon-based, carbon comprises a relatively small component of the Earth's crust (0.032% by weight).
- Carbon occurs in gaseous forms (CO_2, CO, CH_4), dissolved in water (HCO_3, $H2CO_3$), in sediments (detritus and humus) and rock (limestone and fossil fuels), and is incorporated into living tissue (organic molecules). Carbon also occurs in inorganic forms (e.g., graphite and diamonds).

- Carbon fluxes among major ecosystem compartments are given in Figure 4.14. Notice in this diagram of the carbon cycle that the upward pointing arrows for fossil fuel burning and land use changes are not offset by downward pointing arrows. These fluxes result in net increases in carbon to the atmosphere, and are associated with global warming.
- Some carbon is listed as "missing" by scientists. That is, it is known to be released into the atmosphere due to fossil fuel burning and deforestation, but it cannot be accounted for in the atmospheric change in CO_2. It is probably being incorporated into marine and land photosynthesis, but scientists are not sure where.
- Carbon is initially incorporated into the biota via photosynthesis (CO_2 is converted to glucose) (Figure 4.15).
- Some carbon (as glucose) is converted into new plant tissue. Some plant tissue is consumed by animals and converted into animal tissue.
- Carbon is aerobically respired by plants, animals, and microbes and returns to the atmosphere or water as CO_2. Carbon is also returned to the atmosphere via anaerobic respiration and combustion.
- Annually, approximately 15% of the total atmospheric carbon is fixed by photosynthetic plants and respired by plants, animals, and microbes.
- In certain anaerobic conditions such as exist in some wetlands or where decomposition is very slow, such as cold deep sea sediments or polar regions, carbon accumulates, is buried, and forms fossil fuel deposits.
- Humans have altered the carbon cycle through burning fossil fuels and vegetation which releases stored carbon, and through deforestation, which reduces the amount of atmospheric carbon taken up through photosynthesis.
- Although understanding the global carbon cycle is problematic, atmospheric monitoring studies have indicated an increase in atmospheric carbon (a greenhouse gas) over the past 35 years.

What is the carbon-silicate cycle?

- Carbon dioxide readily dissolves in water to form carbonic acid (H_2CO_3).
- As water containing carbonic acid migrates through groundwater and surface water systems, the acid dissolves silicate rock and releases calcium and bicarbonate ions.
- Calcium and bicarbonate ions are used by planktonic organisms and mollusks to construct shells.
- These shells, along with bones and other forms of carbon in sediments, accumulate on the sea floor and form limestone. If this limestone is subducted by tectonic processes, the carbon may be released as CO_2 by volcanic activity (Figure 4.18).

How does the nitrogen cycle function?

- Nitrogen is a constituent of proteins, thus essential for life.
- Free nitrogen (N_2) makes up approximately 80% of the atmosphere.
- Free nitrogen cannot be used directly by plants and animals so must be converted to a usable form. This conversion is accomplished by various species of bacteria. (Lightning also oxidizes N_2 to usable NO_3-.)

- Bacteria living symbiotically with plants (e.g., legumes) and algae, or free living in the soil or water, convert N_2 to ammonia (NH_3) (which can be used by plants); this process is called nitrogen fixation (Figure 4.19).
- Nitrogen-fixing bacteria also live in the stomachs of some animals such as the ruminants and help provide as much as half the nitrogen requirements for the animal.
- When plant and animal tissue or animal wastes decompose, decomposer bacteria convert organic molecules back to usable ammonia (NH_3) or nitrate (NO_3-).
- Under certain anaerobic conditions (such as occur in wetlands), denitrifying bacteria convert nitrate (NO_3-) or nitrite (NO_2-) to free nitrogen; this process is called denitrification.
- Industrial processes can be used to artificially fix nitrogen for use in commercial fertilizers.

How does the phosphorous cycle function?

- Phosphorus does not have a gas phase so the cycle is somewhat simpler than the carbon or nitrogen cycles (Figure 4.20).
- Phosphorus is a component of the DNA and ATP molecules.
- Phosphorus cycles in its oxidized form, phosphate (PO_4), which is taken up directly by plants, algae, and some bacteria.
- One source of phosphate is the waste and remains of animals and plants. Bird and bat guano contains high concentrations of phosphate. Ecosystems containing the feeding grounds, rookeries, and roosts of the animals are often very productive (e.g., ocean upwellings).
- Phosphate is also found in high concentrations in sedimentary rocks containing the fossilized wastes or remains of marine plants and animals. This phosphate can be mined and used in commercial fertilizers.
- Phosphate is often the most limiting nutrient in freshwater ecosystems. Artificial inputs of phosphate into waterways from agricultural, residential, and urban sources can result in eutrophication.

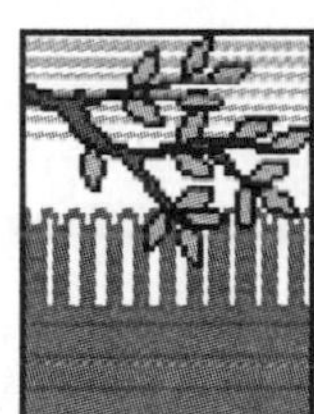

Ecology In Your Backyard

- In many waterways of the United States, eutrophication is a water quality problem. Is eutrophication a problem in the waters in your region? If so, what evidence of eutrophication is apparent? What are the possible sources of excessive nutrients that are causing the problem?
- Some state and local agencies have citizen water quality monitoring programs in which you can participate. Simple water quality test kits used to measure nutrient concentrations and other water quality parameters are usually provided to participants. Although simple field kits are not highly precise, general trends in water quality can be determined.

Links In The Library

- Gilliland, M.W. 1988. A study of the nitrogen-fixing biotechnologies for corn in Mexico. *Environment* 30: 3.
- Kasting, J. F., O. B. Toon, and J. B. Pollack. 1988. How climate evolved on the terrestrial planets. *Scientific American* 258 (2): 90-97.
- Pomeroy, L.R. 1974. *Benchmark Papers in Ecology: Cycles of Essential Elements.* Dowden, Hutchinson and Ross, Inc. Stroudsburg, PA.
- Post, W.M., T. Peng, W. R. Emanual, A.W. King, V. H. Dale, and D. L. DeAngelis. 1990. The global carbon cycle. *American Scientist* 78:310- 326.
- Schlesinger, W. H. 1992. *Biogeochemistry: An Analysis of Global Change.* San Diego: Academic Press.

Ecotest

1. Which of the following is not a macronutrient?
a. phosphorous
b. nitrogen
c. silicon
d. sulfur

2. If a nutrient or element is present in a very low concentration or excessively high or toxic concentrations such that the survival and growth of an organism, population, or species is affected, the nutrient becomes a ______.
a. response variable
b. control parameter
c. flux factor
d. limiting factor

3. During the hydrologic cycle, the energy providing the buoyancy for air and water droplets in clouds comes in part from ____________________.
a. the latent heat of condensation
b. relative humidity
c. saturation

4. Ecosystems that readily exchange a relatively large portion of their nutrients with other ecosystems are called ______ ecosystems.
a. leaky
b. fluctuating
c. dynamic
d. active

5. By weight, what percentage of the earth's crust is carbon?
a. 0.032%
b. 0.32%
c. 3.20%
d. 32.00%

6. When CO_2 dissolves in a freshwater stream, the mild acid ________ is produced, which is capable of slowly dissolving parent rock.
a. H_2CO_3
b. HCO_3
c. $CaCO_3$
d. HCO

7. Nitrogen fixation entails the conversion of ______ by symbiotic and free-living bacteria.
a. N_2 to NH_3
b. NO_3 to NH_3
c. organic molecules to NH_3
d. NH_3 to N_2

8. Denitrification entails the conversion of ______ by anaerobic bacteria living in wetland soils.
a. NO_3 to N_2
b. NH_3 to NO_2
c. NH_4 to N_2
d. N_2 to NO_3

9. Which statement about phosphorous cycling is false?
a. Phosphorous, in the form of phosphate, is taken up directly by plants.
b. Phosphorous is concentrated in marine sediments.
c. Phosphorous gas has a long residence time in the atmosphere.
d. DNA and ATP contain phosphorous.

10. Excessive inputs of nutrients into waterways may result in ______, which is characterized by rapid aquatic plant growth.
a. oligofication
b. eutrophication
c. nutrient fixation
d. nutrient saturation

11. All of the following natural processes are involved in nutrient cycling except:
a. the rock cycle.
b. the tectonic cycle.
c. the hydrologic cycle.
d. the circadian cycle

12. What is the basic structural element of all organic molecules?
a. Boron
b. Calcium
c. Helium
d. Carbon
e. Gold

13. Which statement about the role of ecosystems in biogeochemical cycling is false?
a. Ecosystems are open to nutrient exchange.
b. Nutrient cycling within an ecosystem can occur when decomposition materials from the leaves of a fallen tree are used to help an understory tree grow.
c. The living biota of an ecosystem are capable of cycling other living materials only after they have been reduced to their macro and micronutrients.
d. The energy used to process and cycle nutrients in an ecosystem is derived from within the system.

14. The rock cycle involves:
a. nutrient distribution by the Earth's crust and tectonic plates.
b. weathering of bedrock by a mountain stream.
c. transformation of parent rock material into weathered rock material.
d. igneous rocks being formed by molten lava.
e. All of these are part of the rock cycle.

15. Free nitrogen (N_2) makes up what percent of the Earth's atmosphere?
a. 21%
b. 100%
c. 80%
d. There is no nitrogen in the Earth's atmosphere.

Chapter 5 - THE HUMAN POPULATION AS AN ENVIRONMENTAL PROBLEM

The Big Picture

The human population is growing rapidly. Whereas the total human population on Earth was less than 250,000 people about 11,000 years ago during pre-agricultural times (or as many people as in a small city today), it is estimated that about 6.2 billion people live on Earth in 2002, many of them in developing countries where population growth rates are the greatest. For example, in Bangladesh (see Case Study), each year 2,260,000 people are added in excess of the number that died (the population of 125.7 million is growing by 1.8 % per year). Recently, a hurricane killed 100,000 people in Bangladesh (compare this number with the small number of people killed when a hurricane hits the U.S. coast). Afterwards, the population growth curve for Bangladesh barely showed a fluctuation because of this great natural disaster (Figure 5.2). In fact, the 100,000 people were replaced in just two week's of population growth! This underscores the magnitude of the population growth rate in Bangladesh. All over the world, people are being added to the planet at an average rate of 1.4 % per year, all the while consuming food, producing pollution, and causing changes to the environment. This may not seem like a lot, but the total number of people on Earth increases by about 84,000,000 people per year; that is roughly equivalent to the size of the population of Germany. It has been said that this problem of overpopulation by humans is the greatest environmental problem facing the Earth. Because people in developing countries spend most of their time and money simply meeting their daily food requirements, no one in those countries with high population growth rates will be able to devote any effort to issues of protecting biological diversity, saving endangered species, conserving fisheries, or managing forests. In this chapter, the authors discuss the historical and current human population growth rate, the methods by which population growth is estimated, the impact of exponential growth rate, methods by which population control may be instituted, and finally, the controversy surrounding the methods of population control.

Frequently Asked Questions

What is demography?

- Demography is the study of populations (from Greek, "demos"= population, "graphy" = to describe or draw).
- The demography of any plant or animal population can be done, but the authors concentrate on human demography in this chapter on human demography. Fishery biologists, foresters and conservation biologists all study the demography of particular fishes, trees, or endangered species.

- Because they and many human demographers are interested in population changes over time, it is often said that they study the **population dynamics** of a given species.

What is exponential growth?

- Populations initially grow at exponential rates of increase. Exponential growth occurs when a population increases by a constant percentage each time period (multiplicative increase), rather than by a constant amount (additive increase).
- The series 2, 4, 8, 16, 32, 64, ... is an exponential series that increases by a mutiplicative factor of 2, whereas the series 2, 4, 6, 8, 10, 12... is an arithmetic series that increases by an additive factor of 2.
- Because of the exponential nature of growth, population growth rates are reported as the percentage growth per year.
- Exponential growth is accelerating growth, so that the increases are small at first but large after a while.
- Exponential growth occurs in much the same way as a bank calculates interest on a loan or a savings account, the percentage change from year to year remains constant, but the absolute amount of money that one gains or pays in interest is ever increasing.
- The human population is currently growing exponentially at a rate of 1.5 % per year.

How rapidly is the human population growing?

- The human population for the entire Earth is growing at 1.4% per year, or 86 million people per year.
- There are about 86 million people in the country of Germany, so the Earth gains about one Germany's worth of people per year.
- The Earth's human population had been growing at a rate of 2.1 % during 1965-1970, but has slowed somewhat as population control measures have been implemented.
- For the developing nations, currently there is 1.7 % annual growth or higher (rapid growth).
- For the developed nations, currently there is 1.0 % annual growth or lower (slow growth).

What is the impact of technology and population on the environment?

- Paul Ehrlich, a biologist at Stanford, has proposed an equation that combines the impact of technology as well as population on the environment:

$$T = P \times I$$

- P is the population size of a nation
- I is the impact on the environment of each person in that nation; it is a function of the technology used by the citizens in that nation
- T is the total environmental impact of that population
- The I for a developed nation, like the USA is large; even if there are not many people in such as nation, the total impact T can be large.
- The I for people in a developing nation like Bangladesh is small, but there are many people in such developing nations, so the total impact T can be large
- The total impact of a population is not only a function of population, but of impact per captita, which increases as a nation uses more technology

Who was Thomas Malthus?

- Reverend Thomas Malthus was a British minister, economist, and one of the world's first human demographers (See "A Closer Look" 5.1).
- In 1803, he published "An Essay on the Principle of Populations" in which he detailed the potential for growth in the human population.
- He predicted that the Earth's capability to produce food would be outstripped by human population growth.
- Thus, he was among the first demographers to put forth the concept that the population of humans has a limit to growth.
- His writings influenced the thinking of Charles Darwin, who at the time was developing his theory of evolution "...by Means of Natural Selection", which had at its core the idea that food was limiting for many species, giving rise to new species as they adapted to harsh conditions (see Chapter 7).
- Malthus based his predictions on the population of London, which had been undergoing rapid population growth for some time, but was slowing during his lifetime. He noted increasing levels of homosexuality and sexual promiscuity without procreation (or vice as he called it) as the population grew and he predicted that this would serve in part to slow population growth. But his calculations on human population growth convinced him that this would not be enough to prevent overpopulation ("...the passion between the sexes is necessary and will remain nearly in its present state.."), and that eventually disease and famine would claim large numbers of people.
- So far, in the 200 years since his prediction, the world human population has continued to grow without limit. So, Malthus was wrong. However, see the next question...

Was Malthus right about the limits to human population growth?

- Strictly speaking, he has not been proven correct in most areas of the world, because world food supply has kept pace with the human population.
- The increased food supply has been due to the technological advances of the Green Revolution, which relied on intensive fertilizer and pesticide application to achieve high yields (See Chapter 12).
- In certain areas of the world, especially in Sub-Saharan Africa, the population has risen faster than the food supply, as can be seen in the declining grain production per capita in those regions. So in Sub-Saharan Africa, Malthus was right, but not elsewhere.
- These patterns may change in the future, and many pessimistic human demographers think that Malthus will eventually be proven correct.

What is the logistic growth curve?

- The logistic growth curve is the S-shaped curve typically used in mathematical models of population growth. A plot of this curve shows time (in years, for human populations) on the x-axis versus N, the number of organisms in the population at each time period, on the y-axis (Figure 5.5).

- The population following a logistic growth curve starts out growing nearly at an exponential rate, but then gradually slows as the so-called "carrying capacity" of the environment is reached. This is the maximum number of organisms that can be maintained indefinitely in that environment, and is often shown as an asymptote, K, of such plots.
- The logistic curve has been widely used in modeling populations of fishes, trees, wildlife, and even humans. Demographers use the logistic curve to predict K or the maximum number of organisms that will be in an area. However, very few organisms follow the curve exactly, and the assumptions made by scientists in order to use the model are rarely achieved in nature. It is rare that populations stabilize at a single value of K, but rather they fluctuate as weather, interactions with other species, and random factors in ecosystems change the carrying capacity.
- Nonetheless, human demographers have used the logistic curve to predict the carrying capacity for Earth. Projections made by the UN Population Division depend on which assumptions about changing total fertility rates are used (Figure 5.7).
  - If the slow fertility reduction path is assumed (total fertility rate is assumed to be near the replacement level of 2.1), then the human population will stabilize at 11 billion by the year 2100.
  - If rapid fertility reduction occurs (total fertility rate of 1.6 or below), then the human population will stabilize at 7.7 billion people in 2050, but will begin to decline to 3.6 billion by 2150. It is not known which of these alternatives will occur; in fact, the actual growth is likely to be somewhere in between these two extremes (see the World Bank estimate above).
  - A third possibility is that the human population will continue to grow at a constant path (assuming that the 1998 fertility rate of 2.5 will continue into the next century); this population projection results in an exponential increase beyond 2100. This seems unlikely as well as undesirable (the population would be well over 20 billion by then), but is a possibility.

What is the upper limit (carrying capacity) of the world's human population?

- No one knows for certain. It is unknown for three reasons:
  - 1) the actual number of people on the Earth can only be roughly estimated, (routine censuses are done in only a few countries);
  - 2) the human population has never been this large before, so that every day we are setting new records; and
  - 3) it really depends on the standard of living the people on Earth maintain.
- If all 6.2 billion of us wanted to live like a typical U.S. citizen, the number of people living would be far below where it is now. If we all live like people in developing nations, then we may have many more people.
- Estimates of the number of people that the Earth can support vary from 2.5 billion (this estimate is based on everyone living and eating like a U.S. citizen, which means we have already exceeded the limit) to 40 billion (this assumes all available flat lands, even deserts and colder regions, are turned into farms).
- The World Bank estimates that the human population on Earth will stabilize at 10.1 – 12.5 billion people in the year 2140, but this prediction is based on some assumptions

about female longevity and fertility and the logistic growth curve (see Figure 5.7). However, the logistic growth curve has not yielded correct estimates of the human population growth limit in the past.

How is population growth measured?

- Population growth is measured as a change in the number of people or organisms per unit time (see "Working It Out" 5.1).
- A simple equation to predict the absolute size of population at a later time is to count the number of people alive currently and add to this the net increase in a given time period (say, a year); this can be written as:

$$P_2 = P_1 + (B - D) + (I - E)$$

Where:
P_1 is the number of people alive currently (time 1)
P_2 is the projected number of people alive after a year (time 2)
B is the number of people born during the year
D is the number of people dying during the year
I is the number of people immigrating into the area during the year
E is the number of people emigrating from the area during the year

- Human demographers normally standardize the growth rate as a proportion, or as a percentage, of the total population, or by the number of new people added to a population per 1000 people per year. This is done in order to compare countries or areas with different total numbers of people and to use a year as the standard time unit.
- The following statistics are used to measure change in the population growth rate as a *percentage, proportion* or *crude birth or death rate* (change per 1000 people):

$$\text{birth rate, } b = B/N$$
$$\text{death rate, } d = D/N$$
$$\text{immigration rate, } i = I/N$$
$$\text{emigration rate, } e = E/N$$

where N is the population size

These values can be expressed as proportions (number of births and deaths per person), percentages (number per 100 persons), or as a crude rate (number per 1000 persons). Depending upon the calculation you are asked to do, the exact answer you report may need to be converted from one of these expressions to another, but they will give equivalent results. Be sure to use the same type of expression all through your calculations.

- The annual population growth rate as a proportion of the total population, g, can be calculated as follows:

$$g = (b-d) + (i-e)$$

or

$$g = \left[\left(\frac{B}{N}\right) - \left(\frac{D}{N}\right)\right] + \left[\left(\frac{I}{N}\right) - \left(\frac{E}{N}\right)\right]$$

or, the birth rate minus the death rate plus the immigration rate minus the emigration rate. Remember to multiply g by 100 to convert this to a percentage, and by 1000 to convert it to the number of people per 1000, depending on how you are asked to report it.

- Essentially, this is the same as subtracting the number of people dying or emigrating from an area from the number of new people added through birth or immigration into an area. Thus, death and emigration have the same impact on a population (they lower it), while birth and immigration both raise the population.
- An example:
  - Australia had a population of 18,700,00 people in mid-1998. Australia's 1998 to 1999 birth rate was 261,800/18,700,000 (or 0.014 as a proportion, 1.4 %, or 14 births/1000 people), and its death rate was 130,900/18,700,000 (or 0.007 as a proportion, 07 %, or 7 deaths/1000 people). Assuming that there are an equal number of immigrants and emigrants [thus, (i - e)= 0], what is the population growth rate g?

$$g = \left(\frac{261{,}800 - 130{,}900}{18{,}700{,}000}\right) = 0.007 \text{ as a proportional change}$$

or

$$g = \left(\frac{261{,}800 - 130{,}900}{18{,}700{,}000}\right) \times 100 = 0.7\% \text{ as an annual percentage change}$$

or

$$g = \left(\frac{261{,}800 - 130{,}900}{18{,}700{,}000}\right) \times 1000 = 7 \text{ as a crude change in the number per 1000 people}$$

What is doubling time?

- Doubling time (T) is the amount of time that it takes for an exponentially growing population to double, usually measured in years for human populations.
- Thus, when a population is growing at a rate of 2 % per year, it will take about 35 years to double. There is a simple formula that has been determined empirically that allows this quantity to be calculated:

T = 70/annual *percentage* growth rate for the population

or

$T = 70/g$

with g expressed as a percentage

Remember to convert g to a percentage, if you have not already done so, using the equations in the above section.

- For our Australian example, this works out to:

$$T = 70 / 0.70 = 100 \text{ years}$$

What is the doubling time for the world's human population?

- The Earth's human population will double in about 50 years, according to most recent estimates ($g = 1.4$ % growth per year). This can be calculated as follows:

$$\text{Doubling Time} = T = (70/1.4) = 50 \text{ years}$$

- This is an instantaneous rate, so that the doubling may occur sooner or later depending on how the growth rate changes in the future. But if the growth rate remains constant at 1.4 %, it will take 50 years for the Earth's human population to increase from 5.9 billion to 11.8 billion people.

What is the demographic transition?

- The demographic transition is a pattern of change in birth rates and death rates that has been observed to occur in human populations (Figure 5.8).
- As human societies pass through a series of stages during cultural evolution, from hunter-gatherer societies to industrial age societies, it has been observed that birth rates and death rates are initially high and approximately equal in non-industrial, pre-agricultural human societies (Stage I).
- Death rates begin to decline during the agricultural stage (Stage II) of human society, as improved food availability, medical care, sanitation, the use of soap, and other technologies improve human survival.
- However, during the intermediate stage (Stage II), because of the "passion between the sexes" as Rev. Malthus put it, the birth rate usually remains high after the death rate falls. This leads to a rapid increase in population (birth rate greatly exceeds death rate).
- In Stage III of the demographic transition, birth rates fall to new low levels, approximately equal to death rates, in industrial and post-industrial societies. If birth rates remain low after this transition from Stage II to Stage III, the population will eventually stabilize and stop growing.
- Stage IV may occur in some post-industrial societies, in which chronic diseases, such as heart disease, are cured or long-term survivorship is improved. This would bring about an additional period of population growth.
- It is difficult for many developing nations, such as African and Arabic nations, to lower their birth rates after the death rate has fallen in the intermediate stage. Thus, many countries are still stuck in the intermediate, high-population growth stage (Stage II), because they have high birth rates but low death rates. Population control advocates suggest that these nations must lower their fertility rates by various methods in order to reach ZPG.

What are some of the most rapidly growing countries?

- The Ivory Coast has a population growth rate of 3.8 % per year, which yields a doubling time of 18 years.
- China, the most populous country, is growing at a rate of 1.4 %, and it will double in size in 50 years.

What countries have low population growth rates?

- Belgium, Austria, Great Britain, and Germany grow at about 0.1 % per year, and thus their populations will double in 700 years. Essentially, the populations in these countries are no longer growing very much; they have stabilized.
- In Russia, the crude death rate exceeds the crude birth rate, so that the population is shrinking.
- In the United States, the growth rate was 0.6 % during 1998; thus the doubling time for the U.S. was about 117 years.

What are the potential effects of medical advances on the human population?

- Technological and medical advances (new drugs and therapies, improved sanitation, use of antibiotics and soap, etc.) serve to increase the longevity of the average person.
- Death rates will continue to fall because of these advances.
- Unless birth rates also decline and equal death rates, we will see an increase in population in many countries in the future due to falling death rates.
- Our choices are between stopping medical research (cloning, biotechnology) reducing birth rates through the use of contraception, or allowing Malthusian processes (famine, disease) to limit the human populations. These are all are controversial choices.

How does acute (infectious) and chronic disease affect death rates in a population?

- Acute or infectious diseases appear rapidly, affect many people, then die out, perhaps re appearing later. They include: influenza, the plague, measles, mumps, and cholera.
- Chronic diseases are often not infectious, but occur in a constant low level in the population: they include heart disease, cancer, diabetes, and stroke.
- In a country such as Ecuador in 1987, most of the deaths were due to acute diseases, such as respiratory disease, infections, and parasites (Figure 5.9).
- In the USA in 1987 and 1997, most of the deaths were due to chronic diseases, such as heart disease, cancer, and stroke.
- In the USA in 1900, however, acute diseases were causing more deaths than in 1987
- In terms of deaths due to acute diseases, Ecuador in 1987 is still like the USA was in 1900.

Will AIDS stop population growth?

- Worldwide, there are 33.6 million people with AIDS or HIV, and infectious disease caused by a virus.
- Compare the number of AIDS patients with the estimated growth of the world population each year: a net increase of 86 million people are added each year.

- Even if everyone with HIV and AIDS were to die, AIDS alone will not stop population growth.
- The world's population is growing too rapidly, due to high birth rates.

What are the limiting factors for the human population?

- In the short term (< 1 year), food distribution problems, disease outbreaks and wars can limit population growth.
  - Famines (often caused by droughts and political disturbances) can increase the death rate of a population.
  - Infectious disease outbreaks can limit growth by rapidly increasing the death rate (the plague is an historical example).
  - Nuclear wars could also limit population growth in the short term.
- In the long term (> 10 years), increases in toxins and pollutants, increases in soil erosion, declines in water supplies, disruptions in non-renewable resources (oil), climate change and global warming will increase the death rate, decrease the birth rate or both.

What is the age structure of a population?

- The age structure of a population is the distribution of individuals among different ages (the proportion of individuals in each age class).
- Populations of different countries differ in their age structure. Rapidly growing populations (such as in countries like Kenya; Figure 5.10, left) tend to have many individuals in pre-reproductive age groups, which is less than age 15 in human populations. Some populations have a dominant age group, like the baby boomers in the U.S. (Figure 5.10, center). Other populations that have reached a stable size have an equal number of individuals in all age groups, like in Austria (Figure 5.10, right).
- By constructing a plot of the age distribution of a population, demographers can determine the tendency of a population to grow or stabilize.
- It is also the age distribution that is used for many social and economic calculations, from deciding whether or not to build schools or retirement homes, to determining life insurance premiums.

What is the Total Fertility Rate (TFR) and how does it differ from the Replacement Fertility Rate?

- The **total fertility rate** is the average number of children born during a woman's lifetime.
- The TFR was 3.8 in the U.S. during the post World War II "baby boom" years. In the U.S. now, this rate is approximately 2.0, which is close to the replacement rate.
- A population with a TFR of 3.8 would increase over time, whereas a population with a TFR of 2.0 would decline.
- Bangladesh has a TFR of 3.3 children per female.
- Brazil has a TFR of 2.3 children per female.
- The **replacement fertility rate** is the TFR that exactly replaces a woman and her spouse, on the average. A population at the replacement fertility rate would remain constant in size.

- For humans, the replacement rate is typically around 2.0 children per lifetime, but slightly higher due to the infant mortality rate. It is 2.1 for the USA.
- A population can continue to grow after attaining replacement fertility rate, because of **population momentum**: a population that has many reproductive-aged females, will continue to grow because of time lag effects.

What is population momentum?

- Even after a population has attained the replacement level fertility rate, it will continue to grow for several generations. Because a human generation is 30 years (time between mother and offspring), the effects of a burst of fertility can be felt for many years.
- This tendency for a population to grow even after the population controls have been applied is called **population momentum** or the **population lag effect**.
- The momentum can last anywhere between 50 and 200 years.
- The momentum is due to the effects of age structure.
- For example: women in the U.S. alive today who are part of the baby boom are having children. Even if they each have 2 children (replacement rate), there are so many baby boomer moms that there will be a surplus of children the next generation - a baby boomlet. So, even though the U.S. population has reached the replacement level fertility rate, the population will continue to grow for a while.
- Eventually, if the population maintains the replacement level fertility rate for a long while, the population will stop growing.

What is Zero Population Growth?

- When a population stabilizes, maintains a replacement level fertility rate, or has crude birth rates and death rates approximately equal for a long period of time, it is considered to be at ZPG or Zero Population Growth.
- Although this is a long-term goal for human demographers to achieve worldwide, it has been achieved only in developed countries like the U.S.

How can we stop human population growth?

- There are a number of ways to slow and stop population growth, should we decide to do so:
    - **Delay the age of first reproduction by women.** This will cause a major decrease in the total fertility rate. In nations with rapidly growing populations, most women marry before the age of 20. China's population was growing so rapidly in 1950 that it experienced a major famine. The leaders of China undertook a program of population control, which included laws setting a minimum for marriage ages (18 for women and 20 for men). As a result, the birth rate fell from 32 to 18 per 1000 people, and total fertility went from 5.7 to 2.1 children per female per lifetime.
    - **Educate women.** A natural and painless way to delay the age of first reproduction by women is by encouraging the education of females worldwide. This is the single most important factor in reducing the total fertility rate to replacement rate levels. Women who are educated and work

outside the home have fewer children and make a greater income. There is a clear relationship between total fertility rate and income (Figure 5.11). As women make greater incomes, they have fewer children. And as they have fewer children, they make more money.

- **Encourage women to use breast-feeding**. Women who feed their newborn infants with their breast milk will not ovulate, hence this prevents pregnancy. It is also healthy for the infant, and so this is another natural, painless method of population control.
- **Encourage family planning.** This includes providing greater access to birth control, contraceptives, abortion, and sterilization. These latter two methods are the most controversial from a moral and ethical perspective. However, they are very significant in terms of reducing birth rates. Given the alternative of overpopulation and children dying of starvation after birth rather than before birth, these methods can really be viewed as humane.

Ecology In Your Backyard

- What is the Total Fertility Rate in your area (state or local government)?
- Calculate the Total Fertility Rate for your family lineage. Determine the following: How many children did your mother have? How many children did your grandmothers and great-grandmothers have? If you are typical, the number of children produced by each of these women should have declined as family size decreased during the past generations. If you can, ask your mother and grandmother why she had the number of children she had. Were the births planned?
- How many children do you expect to have? Please state if you are male or female.

Links In The Library

- Cohen, J.E. 1995. *How Many People Can the Earth Support?* W.W. Norton and Company, New York.
- Kent, M. M. and K. A. Crews. 1990. *World Population: Fundamentals of Growth.* Population Reference Bureau, Washington, DC.
- Sadik, N. *The State of World Population. 1990.* United Nations Population Fund Report, New York.
- World Bank, 1985. *Population Change and Economic Development.* Oxford: Oxford University Press.
- Hub, C. 1995. *Global and U.S. National Population Trends*. Consequences, The Nature & Implications of Environmental Change, Volume 1 (2): 3-11.

Ecotest

1. A study of the changes in the size of a population over time is called ______________________.
a. population dynamics
b. population census
c. age-distribution study
d. family planning study

2. Demography refers to:
a. the study of human populations.
b. the study of any population.
c. the census of a population.
d. the age, sex, and racial characteristics of a population of people.

3. What did Malthus predict?
a. The human population will eventually exceed the ability of the Earth to produce food for them.
b. If trends in human population growth are true, we will always be able to feed everyone on Earth.
c. There are no limits to human population growth.
d. Although the population of London stopped growing, the world's population will not.
e. When population density gets high, homosexuality and vice increase

4. Have Malthus's predictions about population growth and food supply been proven correct?
a. Yes. All human populations in all countries have insufficient food.
b. Yes, but not in all areas of the world. Some human populations have enough food, but there have been famines elsewhere.
c. No. Humans populations have grown, but so have food supplies.
d. None of these are correct.

5. At what level will the human population reach a stable upper limit according to the World Bank?
a. 5.7 billion people in 2140
b. 10 - 12 million people in 2000
c. 10 - 12 billion people in 2140
d. 32 billion people in 2000
6. Which of the following series is an exponentially increasing series?
a. 2, 3, 4, 5, 6, 7, 8, ...
b. 2, 4, 6, 8, 10, 12, 14, ...
c. 2, 6, 10, 14, 18, 22, 26, ...

d. 2, 8, 24, 72, 216, 648, 1944, ...
e. None of these are exponentially increasing.

7. If the crude birth rate in Bedsecksistan, a former Soviet republic, is 35 per 1000 and the crude death rate is 30 per 1000, what is the annual growth rate of this population?
a. 0.05 %
b. 0.005 %
c. 0.5 %
d. 5 %
e.- 0.2 %

8. What is the population doubling time for humans in the nearby former Soviet Republic of Gudsecksistan, a country that has a crude birth rate of 40 per 1000 and a crude death rate of 15 per 1000?
a. 14 years
b. 3.5 years
c. 28 years
d. 2800 years

9. Assuming a starting size of 5.7 billion people in 1997, a population growth rate of 1.5 % per year for the human population, and unchanging demographic statistics (total fertility rate, etc.), what will the world population be in the year 2000?
a. approximately 6 billion people
b. 11.4 billion people
c. 5.8 billion people
d. 8 billion people

10. What is the demographic transition?
a. The change in total fertility rate as a country develops from an agricultural to an industrial society.
b. The migration from rural to urban centers of population.
c. A change from high levels of birth and death rates in pre-industrial societies to low levels after industrialization.
d. The change in death rates associated with famine and disease as a country moves into an industrial stage.

11. The following statistics are used to measure population growth rate:
a. Emigration rate (*e*)
b. Crude birth rate (*b*)
c. Immigration rate (*i*)
d. Crude death rate (*d*)
e. All of the above

12. What is the age structure of a population?
a. The distribution of individuals among different ages.
b. A measure of the tendency of a population towards growth.
c. The minimum age before an individual is considered a part of the population.

d. The demographic transition.

13. The tendency for a population to grow even after population controls are applied is called:
a. population boom.
b. population overdrive.
c. population decline.
d. population momentum.
e. population stability.

14. All of the following are methods of stopping or controlling human population growth except:
a. delaying the age of first reproduction.
b. encouraging breast feeding among mothers.
c. migration to outer space and other planets.
d. birth control measures, such as family planning, the pill, sterility programs, and abortion.

15. ZPG stands for and means which of the following?
a. Zero population growth in which no more individuals are being born.
b. Zenith population growth, the point where the population is growing at its fastest.
c. Zero population growth in which a state of equilibrium exists between birth and death rates
d. Zero potential growth, when a system can hold no more individuals.

Chapter 6 - ECOSYSTEMS AND ECOSYSTEM MANAGEMENT

The Big Picture

Managing ecosystems and the populations that live in them requires that humans understand the complexities of those ecosystems, the ecological communities they contain, and interactions among the species in the community. This chapter will explain the concept of the ecosystem and the ecological community in detail. Whereas an ecosystem is composed of all the living and non-living components of the environment in a given area, an ecological community is composed of all the living species (or populations) in a given area that interact with one another. Note that this meaning of "community" is quite different than that normally used by most people, which refers instead to all the people interacting in a social network. Humans attempt to manage species that are endangered or others that have grown so abundant that they have become pests, but without an understanding of how these species fit into their ecosystems and community food webs, humans cannot rationally manage them. Interactions among species in communities and ecosystems are often subtle and a change in one species' abundance will influence many others, often in unpredictable ways. For example, in the Case Study (Figure 6.1), all the species involved in a woodland community food web in the northeastern USA interact with each other; changes in one species' abundance can influence the others. One of these species is the tick *Scapularius* (Figure 6.1c) that carries Lyme disease, a pest species that afflicts humans. The ticks are carried by and feed on mice and deer, and the mice and deer population levels are tied to fluctuations in the oak tree acorn production (they feed on acorns, especially in "mast" years of large acorn production) and the forest clearing patterns created by humans. Thus, the Lyme disease problem is changed by these natural and human-induced changes because of the interactions in a food web. Because the mice also feed on gypsy moth larvae, in years when mice populations decrease because of poor acorn production, there is an increase of gypsy moth larvae, which consume oak tree leaves and can cause the death of the oak forest. Thus, two pest species are involved, and forest and wildlife management policies that reduce mice and deer populations in order to control Lyme disease will also have the side effect (or indirect effect) of increasing gypsy moth larvae. In this chapter, the authors emphasize the complexity of ecosystem and community interactions such as these. In addition, the authors discuss how ecosystems and communities are structured, what trophic levels are and how they function in food chains and food webs, what indirect and direct interactions in ecological communities are, and why ecosystems are difficult to manage, restore and create.

Frequently Asked Questions

What is an Ecosystem?

- An ecosystem is the minimum system of interacting species and their abiotic resources in an area necessary to sustain life. Normally, an ecosystem must have some kind of producers, consumers, and decomposers (see FAQ's on each of these below).
- The boundaries of an ecosystem are often poorly defined or indistinct, because one ecosystem is really not independent of another. Thus, the Atlantic Ocean and the Hudson River are separate ecosystems, but because the Hudson flows into the Atlantic, and water, nutrients, and aquatic organisms are exchanged between them, they are not independent. Boundaries for ecosystem studies must be defined, and the amount of energy and material entering and leaving the defined area of an ecosystem can then be measured.

What is an ecological community?

- An ecological community is composed of all the interacting species in an ecosystem, which are the populations of producers, consumers and decomposers. Because this definition excludes the abiotic environment, it is distinctive from an ecosystem.
- The organisms in a community must interact in some way to be considered part of the same community. The interactions can be direct (a predator consuming prey) or indirect (a predator's prey is outcompeted by another animal, so that both the predator and prey populations are affected. See the FAQ's on direct and indirect interactions below).
- Because ecologists are often unsure if particular species present in an area interact, an operational (practical) definition of a community is all the species that occur in an area. That is, ecologists assume that all species in an area are capable of interacting.
- Communities may be defined in such a way by ecologists that only a portion of the ecological community is examined (i.e., the bird community, the fish community, the plant community, etc.), but as with an ecosystem, the geographic area of the community must also be established.

What does "trophic" mean?

- Trophic comes from the Greek word for nourishment or food. Trophic interactions are feeding interactions depicting which species consumes another.
- Tropho-dynamics are the changes in feeding relationships over space or time in ecosystems.

What is a trophic level?

- A trophic level consists of all the organisms in a community that are the same number of links away from the original source of energy. These organisms are labeled with a number that indicates the number of links or arrows from the sun within the food chain or food web.
- Trophic level numbers are assigned starting with 1 at the base of the chain or web (the plants, primary producers, or autotrophs in the ecosystem are one link away from the sun), 2 for the herbivores (also known as the primary consumers), 3 for the carnivores (or secondary consumers), and continuing until the decomposers are reached.

What is an autotroph?

- An autotroph is an organism that produces its own food (auto = self, trophic = food in Greek). An equivalent term for an autotroph is **producer**. These organisms are always trophic level 1.
- Typical autotrophs in ecosystems are **photoautotrophs**, such as plants, algae, and photosynthetic bacteria, all of which use the energy in sunlight to make the energy-rich molecule glucose, a sugar, from carbon dioxide and water.
- In some marine ecosystems where there is no light (like the deep sea rift vents), there are **chemoautotrophs** at the base of the food web. These chemoautotrophic bacteria make glucose from carbon dioxide and hydrogen sulfide in the water rather than the sun's energy.

What is a heterotroph?

- A heterotroph is an organism that consumes other organisms for food (hetero = other in Greek). An equivalent term for a heterotrophic organism is **consumer**. These organisms maybe at trophic levels 2, 3, 4, 5 or higher.
- All animals are heterotrophic, and many microscopic bacteria, fungi, and protozoans are also heterotrophic. Some algae, especially dinoflagellates, can switch between autotrophy and heterotrophy.
- Decomposers, scavengers, carnivores, herbivores, and omnivores are all different types of heterotrophs.

What is a herbivore?

- A herbivore is a consumer that eats plants or autotrophs. Many insects, rabbits, deer, and cows are examples of herbivores.

What is a carnivore?

- A carnivore is a consumer that kills and eats other consumers, but no producers. Wolves, lions, dogs, and cats are all examples of carnivores.

What is an omnivore?

- An omnivore is a consumer that eats other consumers and producers. Bears, chimpanzees, and humans are all omnivores.

What is a scavenger?

- A scavenger is a non-fungal, non-bacterial consumer that eats recently dead organisms that have been killed by another organism or by an accident. Vultures, beetles, and crabs are scavengers.

What is a decomposer?

- Bacterial and fungal consumers that consume dead organisms are decomposers. Typically, decomposers produce extracellular enzymes to consume their food, whereas scavengers consume the dead organisms and digest them internally.

What is a primary consumer?

- A primary consumer is a herbivore.

What is a secondary consumer?

- A secondary consumer is a carnivore that consumes herbivores.

What is a food chain?

- A food chain is a linear arrangement of feeding relationships that trace energy flow. That is, a food web starts with an autotroph population (the producers), and then a heterotroph population (the primary consumers) consumes some of the autotrophs, and the sequence continues in a chain, with one or more additional heterotroph populations (secondary consumers, tertiary consumers, etc.) included.
- Decomposers are the last heterotrophic populations to consume the energy obtained by the autotrophs at the base of the chain.
- Although this food chain concept is central to understanding ecosystems, there are few isolated, linear food chains in nature. Food chains are actually part of larger, more complex food webs (see next FAQ).

What is a food web?

- It is rare that a species is eaten by only one other species, or that a heterotroph eats only one species, as a food chain implies. A food web is a network of all the possible trophic interactions; thus it is an aggregation of the individual food chains in an ecosystem. Thus, a food web is a more realistic view than a food chain of what happens during trophic interactions in an ecosystem.
- Food webs can be simple (see a Closer Look 6.1, "Yellowstone Hot Springs Food Chain", which is really a food web with 20 species interacting), and relatively complex [see Figure 6.3, "A Terrestrial Food Web" and Figure 6.4 "An Oceanic Food Web"].
- All of the diagrams seen in this book and every other textbook are simplified versions of actual food webs, because at some level species are grouped together into large categories such as "decomposers" or "phytoplankton". Even the complex diagram shown in Figure 6.5 ("The Food Web of the Harp Seal") is a simplified version of an actual food web. It would be desirable to specify the food web structure at the species level, but such a diagram would be an uninterpretable tangle of lines.
- A species in a food web can feed from multiple trophic levels (see the Harp Seal food web Figure 6.5 for an example). In such cases, ecologists assign a species like the harp seal to the trophic level that is one higher than the highest level at which it feeds (trophic level 7).
- Many ecologists are now using computers to analyze food webs. One approach, network analysis, is used to depict the food web structure of an ecosystem. A computer model has no theoretical limit to the number of interactions that can be included, and more importantly, the trophic interaction strengths can be specified (that is, when a species eats more of one species than another, it has a potentially stronger interaction strength on the first species than the second). In such models, intermediate or effective trophic levels can be calculated for each species based on the amount of food that is consumed at each trophic level. So the harp seal, which feeds at multiple trophic levels, would have an effective trophic level of 4.5, because most of its food comes from food chains with four levels. But as can be seem in

Figure 6.5 some of the chains have 3, 5, 6 or 7 levels; the computer model computes the average path length for the food of the harp seal.

How many trophic levels are possible in an ecosystem?

- Some food chains within an ecosystem food web can be shown to have as many as 6 or 7 linkages, so we can say that there may be an upper limit to the number of trophic levels at 7.
- Some small amount of the energy obtained by producers makes it to trophic level 7, but most of it has gone to the decomposers by trophic level 4.
- The decomposers make it difficult to determine how many trophic levels are present, because their identity and diets are poorly studied at the species level. Some of the decomposer's energy is obtained at trophic level 2 (when a plant dies and is decomposed), but some is obtained after the top-most trophic level (such as when a human dies who has eaten sharks, which ate large fishes, which ate smaller fishes, which ate zooplankton, which ate phytoplankton). In this latter case, the decomposer is at trophic level 7.
- Typically it is rare for there to be more than 5 trophic levels in an ecosystem, because about 90 % of the energy is lost from the ecosystem with each trophic transfer, and thus the energy required to support that many trophic transfers is too great (see Chapter 8). If humans were forced to consume nothing but sharks at trophic level 5, for example, we would overfish the world's shark populations within a single year. Mostly humans consume vegetables, grains, or dairy products, which are at lower trophic levels (trophic levels 2 or 3; see Chapter 12).

What is ecosystem and community structure?

- Just as a building has an underlying structure, ecosystems and communities have an underlying framework of interactions among living (or biotic) and non-living (abiotic) parts of the system that are depicted in the food web and the biogeochemical cycles.
- Recall that back in Chapter 4, we learned about the biogeochemical cycles in which interactions occur between abiotic and biotic parts of an ecosystem; these are part of the ecosystem structure, but not the ecological community structure.
- The community structure is restricted to the biotic interactions among species, including trophic interactions, competitive interactions, and symbiotic interactions.
- Ecologists measure ecosystem and community structure using mathematical indices that depend greatly on the number of species present in the community, their interaction strengths in the food web, the interspecific interactions and the amount of abiotic resources present in the ecosystem.

What is a direct trophic interaction?

- When one species kills or consumes another in a community, it usually has a negative effect on the prey population. This is called a direct interaction.
- Much evidence has accumulated that predators can cause direct negative effects on their prey populations, locally causing the extinction of the prey. Unless there is a natural prey refuge or there is a limit by some other species to a predator population, the predator can eliminate the prey everywhere.

- Some species are currently listed as endangered or threatened because the community structure of the ecosystem where they live has been changed by humans and predators are directly causing their extinction. Human hunters are the best known example of this, having caused the extinction of woolly mammoths and saber tooth cats during pre-historic times, the great whales in the recent past, and even today driving Bengal tigers and other wildlife to extinction (see Chapter 12).

What is an indirect trophic interaction?

- An indirect trophic interaction occurs when one species preys upon another and drives its population downward as in a direct effect, but a third species that interacts with the prey is caused to increase (positive indirect interaction) or decrease (negative indirect interaction) in abundance.
- These types of community interactions are very common, but their outcomes are difficult to predict because they require extensive knowledge of the food web structure and interaction strengths among the species.
- The best known example of the indirect community interaction effect occurs in the underwater kelp forests along the coast of the Pacific Northwest of the U.S. as shown in Figure 6.6. In these communities, sea otters feed on sea urchins, which in turn consume the kelp. This simple food chain is really more complex, because humans prey on sea otters, and sea otters can also eat clams and other shellfish. In a typical situation, before humans came to prey heavily on the otters, the kelp forests were abundant. This was because the otters kept the urchin populations in check, and the urchins could only consume a limited amount of kelp. When humans began killing the otters (for fur, mainly) the urchins were no longer kept in check. There was an increase in sea urchin density (positive indirect effect of human predation on otters) around islands where humans lived and caused the local extinction of the sea otters (direct effect by humans preying on otters). However, humans created an additional negative indirect effect with their predation on sea otters: The kelp forests (as measured by percentage of the bottom covered by kelp forest) disappeared in areas without otters. This negative indirect effect occurred because the otters no longer held the urchins in check and they consumed the kelp. In addition, many other species were impacted by this human predation on sea otters, because the kelp forest supported a diversity of fishes and invertebrates that disappeared along with the kelp. It can be seen that changes in trophic interactions can cause far-reaching impacts in ecosystems, and hence much research effort is being devoted right now to understanding food webs and trophic interactions.

What is dominance within an ecological community?

- Dominance is the degree to which one or two species are the most abundant in a community. This will be discussed in greater length in Chapter 7.

What is a keystone species?

- The keystone species is one that, if it is removed or is caused to decrease in abundance, then its absence will cause a large change in community structure.
- The analogy is to the stone that is placed at the top of an arched doorway by masons. If the keystone is removed, the archway will collapse.

- A change in community structure can occur if certain species are removed. The sea otter in the example above is considered to be a keystone species.

What is the holistic or the individualistic view of communities?

- There is some controversy among ecologists over whether communities are holistic or individualistic in nature.
- Those that hold the **holistic view** suggest that communities act like superorganisms, that if any species are removed the community will no longer function as a whole. In support of this idea is the observation that many communities have autotroph, heterotroph, and decomposer species that occur in the same places together all the time. However, this view, while appealing intuitively, cannot be fully supported by experiment and observation. Removal experiments have shown that some species can be easily removed without a major impact on the others. Others species, like the sea otter or just about any dominant autotroph, if removed, can cause a major change in the ecosystem.
- The **individualistic view** suggests that all species occur together by chance and they just happen to survive where they are because they are lucky. Many species (like some trees and coral reef fishes) seem to have a random distribution pattern on a small time or spatial scale, so this lends support to this viewpoint. However, it may turn out that when these species are studied over a broader scale, there will be patterns of co-occurrence with other species that have not yet been detected.
- It is best to think of these viewpoints as representing extremes of a continuum. Some species are more central in the community structure than others and they are ones that give credence to the holistic view. Other species are rare or do well only when natural disturbances have eliminated their competitors, so they support the individualistic view. This is an important theoretical question in ecology, but also one that can have practical implications for saving endangered species and restoring ecosystems damaged by humans. For example, if species can be removed without impacting others greatly, humans may decide to allow species extinctions caused by humans to occur (although nearly every ecologist would oppose such a move no matter which view they hold, because of the danger of losing a species that might be needed later). Alternatively, if all species are needed to keep the Earth's ecosystems alive and functioning, then we should strive at all costs to keep species from going extinct.

What is a watershed?

- Within a watershed, any drop of rain that falls to the ground flows out in the same stream.
- A watershed is a good way to define an ecosystem occurring on land, because much of the nutrients, soils, and water are forced to enter via rainfall and exit that stream or river. The biotic community is less constrained by the watershed boundaries, but often they seem to reside within a single watershed their whole lives.

What is ecosystem management?

- By controlling the inputs and output of certain abiotic factors and removing or planting certain species, humans have learned to manage ecosystems for their benefit.

- Agriculture can be viewed as a form of ecosystem management, because species of crops are planted, fertilizer and water are added, and competitor species (weeds), pest species (herbivorous insects) and predators (like coyotes and wolves) are removed.

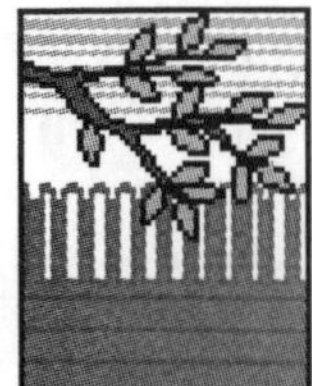

Ecology In Your Backyard

- Can you develop a food web for your own feeding interactions in your ecological community? Try to do this by writing down everything you consume for a week, noting what kind of organism it was made from. Many of the things we eat are made from plants as well as animals and everything we consume comes from an ecosystem somewhere. Take a hamburger, for instance: the meat comes from cattle in the Midwestern U.S. or it is imported from South America, the bun comes from wheat flour, the yeast used to leaven the bun's dough is actually a live fungus and thus a decomposer, the tomatoes and lettuce are grown in California on a farm. Assign a trophic level to each component of the food you consume and you can begin to draw a food web of your diet. Here are some examples of common foods and their trophic levels (remember you would be feeding one trophic level higher) :

Food item	Trophic level
Vegetables	1
Fruits	1
Breads and cereals	1
Rice	1
Pasta without meat sauce	1
Beer or wine	2
Clams, oysters, scallops	2
Pizza (with cheese & vegetables)	2
Pizza (with cheese & meat)	3
Chicken	3
Lobster or shrimp	3
Beef	3
Pork or bacon	3
Cheese	3
Eggs	3
Pasta with meat sauce	3
Fish sandwich (cod, hake, haddock)	4
Fish (tuna, swordfish, shark)	5

Note: Most livestock like pigs, chickens, and cattle are fed prepared feeds that combine grains (trophic level 1) and fish meal (trophic level 3). Thus, by convention used in your text (place the animal one trophic level above the highest trophic level at which they feed), livestock should be at trophic level 4. However, the reality is that much more grain than fish meal is consumed, so that an "average trophic level" can be calculated from a computer model of the diet of the livestock. A trophic level of 3 is probably more realistic for these animals. However, livestock that are allowed to graze and are not fed intensively with supplemented feeds should be placed at trophic level 2 (this is rare). Likewise pizza is composed of bread (level 1), tomatoes (level 1), and cheese (level 3). An average trophic level for a cheese pizza is 2, but for a meat pizza is 3.

- At what trophic level are you feeding most of the time? To determine this, you can count the number of trophic interactions with each of the trophic levels you've recorded above (50 links to trophic level 1, 20 links to trophic level 2, 10 links to trophic level 3, etc.). The one that has the greatest number of links is where you feed most of the time.
- If you were to weigh your food at each meal, then assign that weight to each trophic level, at which trophic level would you consume most of your food on a weight basis?

Links In The Library

- The journals *Ecology, Ecological Monographs,* and *Ecological Applications* are published by the Ecological Society of America. Each month, they contain many new scientific articles on food webs, trophic interactions, direct and indirect effects, and community and ecosystem studies.
- Krebs, C. J. 1985. *Ecology*. Harper Collins, New York. This is an excellent introductory text on Ecology for the interested student. More in-depth discussions of the topics in this chapter are available here.
- Perry, D.A. 1994. *Forest Ecosystems*. John Hopkins University Press, Baltimore. 649 pp.
- Smith, R.L. 1996. *Ecology and Field Biology*. HarperCollins College Publishers, New York. An introductory ecology textbook. 740+ pp.
- Teal, J. and M. Teal. 1969. *Life and Death of the Salt Marsh*. Ballantine Books, New York.

Ecotest

1. An _______________ is the minimum system of interacting species and their abiotic resources in an area necessary to sustain life.
a. ecological community
b. ecosystem
c. ecological food web
d. ecological food chain

2. A species interaction that causes a decline in the abundance of a species one trophic level below its predator is called ______________________________.
a. an indirect negative interaction
b. an indirect positive interaction
c. a direct negative interaction
d. a direct positive interaction

3. The prefix ____________ comes from the Greek word for food.
a. tropho
b. auto
c. photo
d. chemo
e. hetero

4. Which of the following groups in a food web would be most likely to provide the sugar (which is fructose and sucrose, but these are similar to and can be synthesized from glucose) for a cola soft drink?
a. Autotrophs
b. Primary consumers
c. Secondary consumers
d. Decomposers

5. A manatee that feeds on seagrasses is a ________________.
a. plant eater
b. herbivore
c. primary consumer
d. All of these choices are correct.

6. A wolf that feeds on a rabbit (which eats only grass) is a _____________.
a. primary consumer
b. secondary consumer
c. tertiary consumer
d. herbivore
e. All of these choices are correct.

7. Carefully examine the diagram in Figure 6.5 "The Food Web of the Harp Seal". How many trophic levels are involved in the longest food chain that you can locate in this food web? Be sure to start with phytoplankton and end with the harp seal, using the arrows to represent a one-way link between each species (do not consider any decomposer links).
a. 3
b. 4
c. 5
d. 6
e. 7

8. The mushroom on your pizza is an example of a __________.
a. producer
b. herbivore

c. decomposer
d. carnivore
e. None of these are correct

9. The structure of an ecological community is determined by:
a. The interactions among the biotic parts of the community.
b. The interactions among the abiotic and the biotic parts of the community.
c. The interactions among the abiotic parts of the community.
d. None of these are correct.
e. All of these are correct.

10. A keystone species is one that:
a. causes a direct trophic interaction within a community.
b. causes an indirect trophic interaction within a community.
c. causes a major change in the community structure if it is removed.
d. is the same as the dominant species in a community.
e. None of these are correct.

11. Another name for an autotroph is ____________.
a. producer
b. consumer
c. developer
d. scavenger

12. Which statement about scavengers is not true?
a. It cannot be an insect.
b. It cannot be a bacterium.
c. It cannot be a mammal.
d. It eats recently dead organisms.

13. Decomposers differ from scavengers mainly because:
a. they are smaller in size.
b. they do not actively seek out dead material.
c. they do not internally digest their food.
d. they get no nutritional benefit from the dead material.

14. What is dominance within an ecological community?
a. A measure of what organism would win in competition.
b. The degree to which a species is most abundant.
c. How long a species has existed.
d. The genetic makeup of an organism.

15. A predator-prey interaction that causes the abundance of another predator or prey to change is:
a) an indirect trophic interaction.
b) a direct trophic interaction.
c) called scavenging.
d) called secondary consumption

Chapter 7- BIOLOGICAL DIVERSITY

The Big Picture

The concern for global biodiversity has only recently begun to move toward center stage as an international issue. Biodiversity conservation was a central theme at the U.N. Conference on Environment and Development, Rio de Janeiro, 1992 (The Earth Summit). Part of the delay in recognizing the problems facing global biodiversity stems from the fact that relatively little is known about global and, in many cases, even local biodiversity. The number of species of plants, animals, fungi, protists and other microbes that have been taxonomically classified is approximately 1.4 million. But, many eminent scientists who have spent most of their careers studying global biodiversity agree that there are vastly more species that have not been identified; some estimates range as high as 100 million species. Species diversity is only one aspect of biodiversity. Genetic diversity and ecosystem diversity as well as species diversity are components of biodiversity. The diversity of life is a consequence of over 4 billion years of evolution. The winnowing process of natural selection has selected for certain characteristics that allowed species to survive in a given environment, while eliminating those characteristics that were not as favorable. Through evolutionary time, species extinction has been commonplace. In recent years, however, species extinction rates have been greatly accelerated by human activities, primarily land alteration and ecosystem fragmentation (see Case Study on the Shrinking Mississippi Delta), pollution, and over harvesting. The science (and art, as some propose) of restoration ecology is still in an early developmental stage (See the Case Study and Chapter 10). No amount of biodiversity restoration can keep pace with the losses that are now occurring. But, through a combination of preservation, conservation and restoration partial recovery of damaged ecosystems may be possible.

Frequently Asked Questions

What is biodiversity?

- The word "biodiversity" has only been in usage for a short time. Coined by E.O. Wilson in 1982, "biodiversity" has become an often-used term within the scientific community and beyond.
- When people speak of conservation of biological diversity, they can mean the protection of habitat diversity, genetic diversity, the overall number of species, the relative abundances of species, or protecting specific endangered species. Each of these is distinct aspect of biodiversity.

- Biodiversity is an expansion of the concept of species diversity by incorporation of diversity both above and below the species level. Thus biodiversity has three components:
    - **Habitat diversity** is a measure of the different habitats present in an area.
    - **Species diversity** or species richness is a measure of the number of different species in a community.
        - **Species Richness** is the total number of species.
        - **Species evenness** is the relative abundance of species.
        - **Species dominance** is the most abundant species.
    - **Genetic diversity** is a measuring of genetic variation within species populations.
- Biodiversity (and all of biology, for that matter) must be conceptualized within an evolutionary framework. Evolution is the common thread of the natural sciences.

How do species evolve?

- This is one of the most fundamental questions of biology.
- Evolution is the change in the genetic characteristic of a gene pool over time. A *gene pool* is the total genetic information within a population. Changes (variation) in gene pools come about by a number of reasons:
    - **Mutations** which occur at the gene and chromosome level. These are changes in the DNA (deoxyribonucleic acid) chemical bases that occur due to chance events or mutagenic agents (X-rays, toxins, etc.). If a mutation results in a change in a group of individuals that prevents their mating with the parental species, a new species has evolved.
    - **Natural Selection:** the variation in the genetics of individuals provides the raw material for the agents of natural selection (predators, parasites, climate, etc.). An individual whose genetic traits fit them better to the environment will produce relatively more offspring than another less well fit individual. Offspring with the beneficial genetic traits will be represented in future generations, whereas the less well adapted individuals not be represented, or they will be less common.
    - **Migration:** Populations migrate into new habitats with new selective pressures, which can lead to new mutations and natural selection.
    - **Genetic drift:** Changes in a gene pool that can lead to species extinction can result from chance alone. This process is called genetic drift, which is most significantly expressed in small, isolated populations. If a population becomes isolated (or a species is so rare that it becomes endangered), only a subset of the genetic diversity may be present in the population. Then, if the environment changes, and the gene that would allow the species to survive is not present in the small isolated population, that species will go extinct.
- Species may be defined as "a population whose members are able to interbreed freely under natural conditions" (E.O. Wilson, *The Diversity of Life*). Each species on Earth is given a unique scientific name, which is written as two Latin words: the genus (for human, this is *Homo*) and the species (for humans, this is *sapiens*).

- It is important to understand that selective pressures change as environmental conditions change. For example, the introduction of new predators or changes in climatic conditions exert new selective pressures and drive the process of natural selection.
- Biological and evolutionary success is *getting one's genes into the next generation.* This is how species win at the "game of evolution."

What is an example of natural selection?

- Malaria is a disease that affects 1.1 million people per year and is caused by mosquitoes, genus *Anopheles,* that carry a parasite, genus *Plasmodium* (See a Closer Look 7.1).
- It was discovered that DDT kills the mosquitoes and quinine kills the parasite; widespread treatments with these toxins in the 1950's and 1960's reduced the prevalence of malaria in 80% of the targeted areas. This was regarded as a great success in medicine.
- But, the disease returned in the 1970's. Why? Resistant genetic varieties of the mosquitoes (resistant to DDT) and the parasite (resistant to quinine) increased as a percentage of the populations. The DDT and quinine had acted as agents of natural selection, but did not eliminate all the disease-causing organisms. The remaining offspring of the mosquitoes and the parasites were genetically resistant to DDT and quinine, and thus more fit to reproduce in the new environment. Quinine resistant *Plasmodium* has been reported in 80% of the 92 countries where malaria occurs.
- DDT has been banned in the USA, because it is toxic to humans and wildlife. New drugs are being developed for malaria, but in the future natural selection may make the parasite resistant to these as well.

How can the history of life be condensed into usable framework?

- Biologists and geologists use fossils from the geologic record to categorize the major events in the evolution of life by date.
- The oldest fossils are bacteria (similar to the Archaea bacteria of today) that have been dated as 3.5 billion years old.
- For 2 billion years after that, only anaerobic microorganisms lived on Earth.
- Photosynthetic organisms developed later, which released oxygen into the atmosphere, allowing aerobic life forms, including the vertebrates and humans, to evolve.

How is species diversity measured?

- **Species richness** is a count of the total number of species in the area under study (Figure 7.3). Richness is expressed as a single number.
- **Species evenness** is a measure of the relative abundance of the species in the area under study. Evenness is expressed as a percentage.
- **Species dominance** is a more difficult concept. Dominance may be measured in several ways: numerical dominance, biomass, coverage, ability to amass resources, or a combination of parameters, such as species relative abundance, relative frequency (number of times a species is encountered in a survey), and coverage.

- There are several species diversity indices (mathematical equations such as the Shannon Index) used by ecologists, but alone they will not provide an adequate assessment of diversity. Species diversity should be expressed in combination with species richness and evenness, as well as the Shannon Index.

How many species are there?
- The number of species that have been described taxonomically (given a genus and species name) stands *very roughly* at 1.4 million.
- Far more species have not been taxonomically described. Projections by several well-known and respected scientists estimate that the number of undescribed or unknown species ranges to 1.4 million or 100 million (Table 7.1).
- Certainly some taxonomic groups are well-described and well known. The number of bird species and the number of mammal species are quite precise; few new species in these groups are likely to be discovered.
- There are three main domains of organisms: the Eukaryotes (all animals, plants, protests or single-celled organisms, and fungi); Bacteria; and the Archaea.
- Other groups, such as insects, the most species rich taxonomic group with over 750,000 known species, are still poorly described. The least described and most difficult to classify are microbes, of which the species diversity cannot be estimated with any reasonable degree of reliability.

How do species interact?
- Interactions between individuals of the same species are called **intraspecific interactions**. Intraspecific interactions include competing for food, mates, space, and territories, as well as mutual defense and other cooperative behaviors.
- Interactions between individuals belonging to different species are called **interspecific interactions** (Table 7.2). Interspecific interactions include competition for resources (e.g., food, water, and space), symbiosis (i.e., mutualism), and predation-parasitism.
- Both intraspecific and interspecific interactions play major roles in natural selection evolution and overall diversity. Intraspecific and interspecific interactions are central themes in the science of ecology.

What is the competitive exclusion principle?
- When two species compete for the same resources, the resources become depleted and both species are affected detrimentally. However, no two species exploit resources quite the same way; one species will be able to exploit the available resources better than the other.
- This is basis for the **competitive exclusion principle**: *two species that have the same requirements cannot coexist in exactly the same habitat.* Thus, one species will exclude the other.
- Some examples of competitive exclusion are: the gray squirrel was introduced into England, which is displacing the native red squirrel (Figure 7.5); and two species of flour beetles, which drive one another to extinction in grain silos where they co-occur, depending on the temperature and moisture conditions (Figure 7.6).

How do species coexist?

- Species coexist by exploiting the available resources in a sufficiently different manner such that competition is avoided or reduced and coexistence is possible.
- Thus, each species has a different **ecological niche**, which may be defined as the species' functional position or ecological role in its habitat (how it interrelates with its environment and other species within its environment).
- By each species having a separate niche, many species are able to coexist and share resources in the same community.

How are niches measured?

- The niche occupied by a species may be quantified if the environmental conditions under which the species lives and the resource base of the species are known.
- For simplicity, one environmental parameter can be measured at a time, for example, the temperature range in which a species (a freshwater flatworm) survives (Figure 7.7). In the case of flatworm "a", if no competing species are present, the worm has a relatively broad *niche width*; this is its **fundamental niche**. In nature, however, other species may be better competitors under certain conditions. The niche of flatworm "a" is reduced in size by the presence of a competing species, flatworm "b". This reduced niche width is the **realized niche**.
- The ecological niche of a species cannot be adequately determined by a single parameter, all of a species' requirements must be accounted for.
- Some species have broad niche widths. These species are *generalists*, capable of exploiting a variety of resources in a variety of ways. A good generalist species is the coyote, a carnivore that eats insects, birds, reptile, mammals, and carrion, and lives in a great variety of habitats.
- Some species, *specialists*, have narrow niches; that is they are highly specialized to exploit their environment in a very precise manner. Specialists have few options if changes occur in their environment. The ivory-billed woodpecker was a specialist. This large bird required old-growth swamp forests in the southeastern U.S. Most of these swamp forests were logged a century ago; consequently the ivory-billed woodpecker is now extinct.

What is symbiosis?

- **Symbiosis** means living together. Many symbiotic relationships are beneficial to both symbionts, this is called *mutualism.* Parasitism is also a symbiotic relationship but the host is affected detrimentally.
- Examples of mutualistic symbiosis are commonplace. Bees pollinating flowers benefit both species; specialized fungi living in association with plant roots and facilitating nitrogen-fixation; and microbes (microflora) living in the gut of **ruminants** and other animals (including humans) facilitate digestion and assimilation (Figure 7.8).
- Symbiotic relationships exist because the symbionts coevolved. *Coevolution* is the complementary evolution of two or more species, each species affecting the

evolutionary course of the other. Coevolution occurs in predator-prey relationships and well as symbiotic relationships.

How does predation affect diversity?

- Predators exert a strong influence on the abundance of prey populations. Predators keep prey populations in check, thereby limiting the size of the prey gene pool and preventing a single or a few prey species from dominating the community.
- In many predator-prey relationships, predators selectively remove the poorly adapted individuals from the prey population, thus driving natural selection.
- Conversely, the size of the prey population influences natural selection in predator populations. If prey populations are low, only the best-adapted predators will survive.

Why is biodiversity high in some locations but low in others?

- In general, low latitude environments (tropical) are higher in biodiversity than higher latitude environments (more temperate or polar). Similarly, low elevation environments have a higher diversity than high elevation environments.
- Altitudinal *ecological gradients* resemble latitudinal gradients (Figure 7.10). For example, the base of Mt. Kenya is situated in a tropical savanna but the top of the mountain is tundra and glacier, even though the latitude is less than 50 km south of the equator.
- Although species richness may be low in high latitude areas, the abundance of individual species can be high. For example, in the world's most productive commercial fishing ground in mid-to-high latitude oceanic regions, the catch may consist of only a handful of species but many tons of those species. Whereas a tropical coral reef has numerous fish species but no single species is in great abundance.
- There are many theories to explain global biodiversity. These theories have been condensed and summarized into a set of factors listed in Table 7.3.
- In general, biodiversity is highest in areas of complex habitat (e.g., a tropical rain forest with multiple forest strata or layers), high productivity, stable climate, moderate disturbance, mid-succession, and where evolution has been allowed to proceed uninterrupted by major climatic or geologic upheavals for a long period of time.
- Biodiversity is lowest in areas of extreme stress, extreme physical environment, severely limited resources, frequent and/or intense disturbance, geographic isolation, and where exotic species have been introduced.

Ecology In Your Backyard

- **Calculate the Shannon diversity index and "species" richness for cars in a parking lot.** Diversity indices are typically used to measure species diversity in nature, but, just for fun, you could calculate a "diversity index" for cars in a parking lot!

- The calculations of species diversity, evenness, and richness are the same (see equations and example data set below).
- This kind of inter-community comparison is done by ecologists attempting to assess the impacts of human development or pollution. When pollution is present or a human disturbance has occurred in an ecological community, diversity is typically low.
- **Example of calculating species diversity in two parking lot "communities"**:
- Count the number of individuals of different "species" of cars in each parking lot "community".
- The number of "species" or car types in each lot = S
- First, calculate **relative abundance (p_i)** for each "species" in each "community" (i.e., the proportion of the "community" represented by each "species" i):

$$p_i = \frac{n_i}{N}$$

Where:

n_i = number of "individuals" in "species" i,

and

N = total number of "individuals" of all "species"

- Next calculate the **Shannon Diversity Index** :

$$H' = -\sum_{i=1}^{S}\left[p_i \times \left(\ln(p_i)\right)\right]$$

- The summation sign $\sum_{i=1}^{S}[\ ]$ means that you should perform the operations inside the $[\ ]$ for each "species" i, starting with "species" i = 1. Add the result inside the $[\ ]$ for the first "species" to the result inside the $[\ ]$ for the next "species", and so on , and stop when you get to the last "species", i = S.
- Note that there is a negative sign before the summation sign (Σ), which means that your answers will always be positive, because the quantity $\sum_{i=1}^{S}\left[p_i \times \left(\ln(p_i)\right)\right]$ will always be negative.

Here's an example data set:

"Species" of car*	"Species" identifier code i	Number of "individuals" in Parking lot A n_i	p_i	$\ln(p_i)$	$p_i(\ln(p_i))$
Chrysler Lebaron	1	10	0.17	-1.7719	-0.3012
Dodge Minivan	2	10	0.17	-1.7719	-0.3012
Toyota Corolla	3	10	0.17	-1.7719	-0.3012
Chevy Cavalier	4	10	0.17	-1.7719	-0.3012
Nissan Pickup	5	10	0.17	-1.7719	-0.3012
Ford Taurus	6	10	0.17	-1.7719	-0.3012
TOTAL	**S= 6**	**N = 60**	**1.00**		**H' = 1.8074**

"Species" of car*	"Species" identifier code i	Number of "individuals" in Parking lot B n_i	p_i	$\ln(p_i)$	$p_i(\ln(p_i))$
Chrysler Lebaron	1	1	0.02	-3.9120	-0.0782
Dodge Minivan	2	2	0.03	-3.5066	-0.1052
Toyota Corolla	3	25	0.42	-0.8675	-0.3644
Chevy Cavalier	4	32	0.53	-0.6348	-0.3364
Nissan Pickup	5	0	0.00	0	0
Ford Taurus	6	0	0.00	0	0
TOTAL	**S= 4**	**60**	**1.00**		**H' = 0.8842**

*Note: Do not worry if the car "species" list differs somewhat between "communities"

- **Which parking lot "community" above is most diverse?**
  - Lot A (S= 6, H' = 1.8074) is most diverse. The car "species" are equally represented in this lot. We say that this "community" has a high degree of **evenness.**
  - Lot B is less diverse based on our indexes (S= 4, H'= 0.8842) and has low evenness, because the car "species" are unequally represented.
  - Chevy Cavaliers are the most common "species" in Lot B, followed by Toyota Corollas. This lot has a high degree of **dominance** by these two "species" ($p_3 + p_4 = 0.95$, or 95 % of the individuals in this community are Chevys and Toyotas).
  - These comparisons are summarized below:

Diversity Measurement	Community A	Community B
Species richness (S)	6 species	4 species
Evenness	High	Low
Dominance	Low	High
Overall diversity	High	Low
Shannon Diversity Index (H')	1.8074	0.8842

- **Now, collect your data outside in a real parking lot!**
- Choose a safe parking lot, preferably one without much immigration and emigration (cars coming and going) and record the number of cars by "species", i.e., their manufacturer (e.g., Ford, Chevrolet, Toyota, or Ferrari).
- Diversity indices are used for comparative purposes, so compare your parking lot with another parking lot, for example, a faculty vs. a student parking lot, or compare the same parking lot at different times. Which car "species" is the most common, or dominant "species", in each lot?
- If you compared the student lot to a car dealer's lot, which one would show the greatest diversity in types of cars? The car dealer's lot would be characterized as a low diversity "community", while the student lot would have more "species" of cars, and would thus be considered high diversity.
- *If the campus police look at you suspiciously, show them your data.*
- Remember to count the number of "individual" cars in each "species" and enter that number in the third column on the data sheets that follow.
- Use a calculator with a ln (natural log) function to calculate the remaining two columns, and add up the negative values in the last column. Multiply the sum by -1 and you have calculated H'.
- You may want to group similar vehicles like "Ford pick-up trucks" or "Chevy sedans" into a single "species".
- Do not worry if the car "species" list differs somewhat between "communities".
- Sample at least 100 cars in each of the lots you survey.

"Species" of Cars in Lot A	"Species" identifier i	Number of "individuals" n_i	p_i	$\ln(p_i)$	$p_i(\ln(p_i))$
	1				
	2				
	3				
	4				
	5				
	6				
	7				
	8				
	9				
	10				
	11				
	12				
	13				
	14				
	15				
	16				
	17				
	18				
	19				
	20				
	21				
	22				
	23				
	24				
	25				
	26				
	27				
	28				
	29				
	30				
TOTAL					

"Species" of Cars in Lot B	"Species" identifier i	Number of "individuals" n_i	p_i	$\ln(p_i)$	$p_i(\ln(p_i))$
	1				
	2				
	3				
	4				
	5				
	6				
	7				
	8				
	9				
	10				
	11				
	12				
	13				
	14				
	15				
	16				
	17				
	18				
	19				
	20				
	21				
	22				
	23				
	24				
	25				
	26				
	27				
	28				
	29				
	30				
TOTAL					

- **Birding is an excellent and fun way to learn about biodiversity**. Additionally, birds are good indicators of habitat conditions; many bird species are very specialized with strict habitat requirements.
  - Tracking migratory bird populations is especially important and highlights the need to address environmental issues on an international level. Migratory bird habitats must be maintained over a very broad area. It is not sufficient to maintain habitats at one end of their migratory route and not at the other.
  - You do not need to take a course in ornithology or belong to a club to take up birding. A good pair of binoculars and a field guide to birds constitutes the basic equipment. The National Geographic Society, the Peterson Field Guide series, and the National Audubon Society series produce popular field guides for North American birds. Keep a journal of your observations.
  - Calculate the species diversity of the bird community in your local area. Compare diversity in different areas.

Links In The Library

- Darwin, C. 1859. *The Origin of Species*. Murray, London.
- Dawkins, R. 1989. *The Selfish Gene*. Oxford University Press, New York.
- Gould, S.J. (ed.). 1993. *The Book of Life*. W.W. Norton, New York.
- Hilty, S. 1994. *Birds of Tropical America: A Watchers Introduction to Behavior, Breeding and Diversity*. Chapters Publishing Company, Shelburne, Vermont. 304 pp.
- Rosenzweig, M.L. 1995. *Species Diversity in Time and Space*. Cambridge University Press, New York.
- U.S. Fish and Wildlife Service. *A Community Profile* series published as FWS/OBS-documents published in the 1980s -1990s covering a broad range of ecosystems.
- Ward, P. 1994. *The End of Evolution: A Journey in Search of Clues to the Third Mass Extinction Facing Planet Earth*. Bantam Books, New York. 301 pp.
- Wilson, E.O. 1992. *The Diversity of Life*. W.W. Norton Company, New York. 424 pp.

Ecotest

1. "Biodiversity" incorporates the diversity of life at three levels; which one of the following is not one of those levels?
a. Species diversity.
b. Genetic diversity.
c. Habitat diversity.
d. Trophic diversity.

2. When people talk about conservation of biodiversity, they can mean protecting:
a. a variety of habitats.
b. the number of genetic varieties.
c. the number of species.
d. individual endangered species.
e. all of the above

3. Changes that occur over time in the gene pool of a population constitute _______.
a. evolution
b. a species' niche
c. the genetic diversity of a species
d. species interactions

4. All of the following are tenets of natural selection except:
a. ecosystem equilibrium must be maintained.
b. genetic variability exists within populations.
c. selective pressures are imposed by the current environment.
d. differential survival and reproduction influences future gene pools.

5. Which of the following is an example of natural selection occurring?
a. The gray squirrel displacing the red squirrel in England.
b. Malaria was nearly eliminated by the widespread use of DDT and quinine, but it began to reoccur as mosquitoes and a parasite became resistant to these toxins.
c. The coexistence of two different types of beetles in a flour silo.

6. Species ______ is a count of the number of species in a given area.
a. dominance
b. frequency
c. richness
d. evenness

7. Species ______ is a the measure of the relative abundance of a species in a given area.
a. dominance
b. frequency
c. richness
d. evenness

8. The ________ principle states that two species that have the same requirements cannot coexist in the same habitat.
a. competitive exclusion
b. niche inhibition
c. intraspecific competition
d. realized niche

9. Ecosystems with high biodiversity tend to have one or more of the following characteristics, except _______, which is a characteristic of low biodiversity ecosystems.
a. stable climate
b. intermediate frequency and intensity of disturbance
c. complex habitat
d. high latitude

10. Based on the fossil evidence, how old is life on Earth?
a. 4000 years
b. 10 million years
c. 2 billion years
d. 3.5 billion years
e. None of these are correct

11. What process was responsible for the change of the DDT-resistance of malaria-carrying mosquitoes (*Anopheles*) ?
a. Industrial pollution caused a mutation in the DNA of the mosquitoes, causing them to become resistant.
b. A DDT-resistant mutation was already present in some of the mosquitoes, and the offspring of these resistant mosquitoes survived and reproduced more frequently than the non-resistant strain.
c. The mosquitoes carried a parasite that conferred resistance to DDT in their DNA.

12. The number of species that have been described taxonomically stands roughly at:
a. 1.4 billion
b. 14,000
c. 500,000
d. 1.4 million

13. Interactions between individuals of the same species are called:
a. interspecific interactions.
b. indirect trophic interactions.
c. extrinsic interactions.
d. intraspecific interactions.

14. Species diversity can be adequately assessed by which of the following measures?
a. species richness
b. the Shannon diversity index
c. species evenness
d. All of the above (none of the above alone can adequately assess the diversity of an area).

15. Which community has higher species richness?
a. a pet store aquarium with 1000 goldfish in it.
b. a farm pond with bass, pickerel, and sunfish in it.

Chapter 8 – BIOGEOGRAPHY

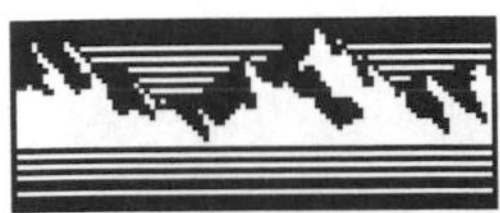

The Big Picture

The Earth's Biosphere can be divided into smaller regions, each containing a characteristic flora and fauna; the study of geographic patterns of species in ecology is called **biogeography**. The species living in such regions all interact and have been subjected to a common geological, climatic and evolutionary history. The name we give such regions is a biome; these are the largest distinguishable subregions of the biosphere with similar climate and species characteristics. Examples on land include deserts, rainforests, grasslands, and forests. There are many different levels of subdivisions that can be used to characterize such biomes, each of which can be modeled as ecosystems. At the lowest level of biological organization in ecology, we can also define species biogeography; each species has a particular biogeographic range over which it can disperse and live. Some species have greatly expanded their range, with the help of humans and our ability to carry flora and fauna with us when we travel. For example, a European species, the purple loosestrife, which is a flowering marsh plant, has caused many problems for the biodiversity of plants in the USA (see Case Study). Such introduced species (or exotic species) can out-compete the native species and often have no natural predators in the areas where they are introduced. Consequently they reduce biodiversity in areas where they gain a foothold. The theory of island biogeography is particularly important in protecting endangered species in conservation biology (see Chapter 14) , because habitats are like islands in a sea of developed and disturbed land.

Frequently Asked Questions

What is biogeography?

- **Biogeography** examines the global distribution patterns of living things.
- Biogeography, as a scientific endeavor, came into its own during the nineteenth century when European explorers traveled the globe cataloging natural history observations for scientific as well as economic reasons. Alexander von Humboldt, whom the Humboldt Current and Humboldt County, California are named after, and Andreas Schimper, who coined the term "rain forest", are but two of these explorers.
- Biogeographers use any of several classification schemes to categorize global biodiversity.

- **Biogeographic realms** are based upon evolutionary relationship and morphological similarities of animals (Figure 8.2). Alfred Wallace divided the world into six biogeographic realms: the Nearctic, Neotropical, Palearctic, Ethiopian, Oriental, and Australian.
- The concept of the **biotic province** is an expansion of the biogeographic realm to include plants (Figure 8.3).

What are exotic species?

- **Exotic species**, or non-native species, are species that have been introduced into a new geographical area.
- **Endemic species**, or native species, are species native to a specific area.
- The term "exotic species" has often been misused to describe species with striking appearances or unusual behaviors. For example, toucans are certainly striking in appearance but they are not exotic to the forests of Central America, they are endemic. Toucans that escape from captivity in south Florida are truly exotic.
- In many instances, exotic species become pest species by outcompeting or preying upon endemic species. Hawaii, Florida and California have experienced major ecological problems with exotic plants and animals that have been introduced by humans.
- **Cosmopolitan species** are found worldwide, wherever the appropriate habitat exists.
- **Ubiquitous species** are found almost everywhere, worldwide, in a range of habitats.

What are convergent evolution and divergent evolution?

- **Convergent evolution** is the evolution of similar morphologies in unrelated (or distantly related) organisms (Fig. 8.7). Convergent evolution occurs as different species adapt to similar environmental pressures. For example, fish, whales, and ichthyosaurs (aquatic dinosaurs) each evolved similar morphologies (fins and tails) for swimming.
- **Divergent evolution** is the evolution of new species from a common ancestor (Figure 8.9). Geographic isolation (and subsequent genetic drift) is a primary cause of divergence. It is assumed that all bird species diverged from a common ancestor, *Archaeopteryx*.

What is island biogeography?

- Islands may be considered microcosms, from which studies of global biogeography, biodiversity and evolution may be extrapolated.
- Ecological islands are not only bits of land in the ocean, but any isolated habitat could be considered an island. For example, in the Southwest, conifer forests on isolated mountaintops are isolated from other mountaintop conifer forests by desert valleys.
- Islands have fewer species than mainlands, and islands have fewer resources than mainlands.
- Islands have high endemism (unique species). One reason for this is **adaptive radiation**. In order to reduce competition for the limited resources on islands, new species diverged from ancestral species to fill unoccupied niches. The adaptive radiation of Darwin's finches in the Galapagos Islands or honeycreepers in Hawaii (Figure 8.10) are often-cited examples. In terms of species diversity on islands, two

aspects of islands are most important: island size and isolation (Figures 8.11 and 8.12).

- Based on studies of island biogeography, the following generalizations may be made:
    - Small islands have lower species diversity than large islands. Small islands are small targets for potential colonists and small islands can support a limited number of individuals. Extinction on small islands is more likely.
    - Distant islands have lower species diversity than nearby islands. Distant islands are more difficult to colonize than nearby islands.
    - Species richness on islands is constant. The species composition may change as one species replaces another but the total numbers of species stays the same. New arrivals either will not become successfully established or they will replace species already on the island.
- The lessons learned from studying biogeography are very important if we are to develop a better understanding of global biodiversity, species extinction, and changing landscapes.

The fragmentation of the landscape by agriculture, forestry, and urbanization creates ecological islands. Many species that require large tracts of unaltered habitat cannot survive in small patches, thus face extinction.

What are Biomes?

- **Biomes** (for a world biome map, see Figure 8.13) are based upon the growth forms of vegetation in response to climatic conditions (Figure 8.14). Climatic conditions (temperature and moisture regime) determine plant physiology and morphology (Figures 8.6 and 8.15). There are at least 17 major biomes (according to your author): tundra, taiga (boreal forests), temperate forests, temperate rain forests, temperate woodlands, temperate shrub lands, temperate grasslands, tropical rain forests, tropical seasonal forests and savannas, desert, wetlands, freshwaters, intertidal areas, open oceans, benthos, upwellings, and hydrothermal vents.

TERRESTRIAL BIOMES	DISTINGUISHING FEATURES
Tundra	Cold climate; high latitude (arctic tundra) or high elevation (alpine tundra); grasses, sedges, dwarf herbaceous species; treeless; year-round freeze potential.
Taiga or Boreal Forests	Cold winters, cool summers; dense stands of conifers (spruce, fir, and larch); low tree species richness.
Temperate Forests	Moderate climate; deciduous trees are dominant; moderate tree species richness.
Temperate Rain Forests	Moist, cool climate; large conifers are dominant; low species richness.
Temperate Woodlands	Moderately dry climate; open canopy; well-lit understory.
Temperate Shrublands	Dry summers, cool winters with low rainfall; low-stature, shrubby vegetation; frequent fires.
Temperate Grasslands	Dry to moderately dry; too dry to support trees; grasses with extensive root systems are dominant; deep organic soils; intensively grazed and frequently burned.

Tropical Rain Forests	Warm, moist or wet year round, large trees; highest tree species richness; numerous epiphytes, vines, and climbing plants; high diversity of all major life forms.
Tropical Seasonal Forests and Savannas	Distinct wet and dry seasons; monsoonal climate; open woodland (seasonal forest) or grassland (savanna); supports large number of grazing animals.
Deserts	Extremely dry, warm or cool; plants are widely spaced and have specialized morphology and physiology to reduce moisture loss; many reptiles are well adapted to desert environment.
AQUATIC AND MARINE SYSTEMS	**DISTINGUISHING FEATURES**
Wetlands	Saturated soils or shallow water; anaerobic, often organic soils; wetland-adapted plants ranging from hardwood forests to grasses.
Fresh Waters	Flowing water (lotic) or still water (lentic) ecosystems; complex food webs based on phytoplankton, macrophytes (large aquatic plants), or allocthonous organic debris (such as leaves falling into water).
Intertidal	Harsh environment due to wave action and alternating exposure and inundation by saline water; algae and mollusks attached on rocky substrates; on sandy substrates burrowing animals dominate.
Open Ocean	Nutrient-poor, low productivity per unit area; pelagic organisms occur at low densities; global productivity is high due to vastness of surface area.
Benthos	Bottom zone; little or no photosynthesis; food chain is supported by detritus drifting down from above.
Upwellings	Nutrient-rich water is carried to the surface; stimulated phytoplankton growth and supports productive food web; important commercial fishing grounds.
Hydrothermal Vents	Hot, sulfur-rich water from tectonic vents stimulates growth of chemosynthetic bacteria that are the basis of food chain.

Ecology In Your Backyard

The next time you travel, record the species that you observe (you will have to leave your backyard for this exercise). What biome are you in after traveling to your new destination? Is this biome the same as where you started? (If you did not travel far, say across town, you will likely be in the same biome). What climate and weather differences can you note? Check the weather in the paper or on the news; daily temperature highs and lows, rainfall, etc. may be different. What new species can you observe? How are they different (shape, color, size, behavior, etc.) from the species in the biome where you started out? For example, if you travel from North Carolina to New

Mexico, you will notice fewer trees, more lizards, more heat and less water as you travel from a temperate deciduous forest biome (North Carolina) to a desert biome (New Mexico). Because of the different climates (moisture and temperature) in these two areas, there are different communities, ecosystems, and species with different morphology and adaptations.

Links In The Library

- Cole, D. J. A. and G.C. Brander (eds.). 1986. Bioindustrial ecosystems. Series: Ecosystems of the world. Amsterdam ; New York : Elsevier.
- Cox, C. Barry and Peter D. Moore. 2000. Biogeography : an ecological and evolutionary approach. 6th Edition. Oxford, Malden, MA, Blackwell Science.
- Cushing, C.E., K.W. Cummins, G.W. Minshall (eds). 1995. River and stream ecosystems. Series: Ecosystems of the world. Amsterdam; New York : Elsevier.
- di Castri, Francesco, David W. Goodall, and Raymond Specht (eds.). 1981. Mediterranean-type shrublands. Series: Ecosystems of the world. Amsterdam ; New York : Elsevier.
- Dubinsky, Z. 1990. Coral reefs. Series: Ecosystems of the world. Amsterdam ; New York : Elsevier.
- Evenari, Michael, Imanuel Noy-Meir, and David W. Goodall (eds.). 1986. Hot deserts and arid shrublands. Series: Ecosystems of the world. Amsterdam ; New York : Elsevier
- Lugo, Ariel, Mark Brinson, and Sandra Brown (eds.). 1990. Forested wetlands. Series: Ecosystems of the world. Amsterdam; New York : Elsevier.
- MacArthur, Robert H. and Edward O. Wilson. 1967. The Theory of Island Biogeography. Princeton, N.J., Princeton University Press, 1967.
- Maria, José; da Silva, Cardoso; Bates, John M. 2002. Biogeographic Patterns and Conservation in the South American Cerrado: A Tropical Savannah Hotspot. Bioscience, Mar 2002, Vol. 52: 225-234.
- Mathieson, A.C. and P.H. Nienhuis (eds.). 1991. Intertidal and littoral ecosystems. Series: Ecosystems of the world. Amsterdam ; New York : Elsevier.
- Postma, H. and J.J. Zijlstra. 1988. Continental shelves. Series: Ecosystems of the world. Amsterdam ; New York : Elsevier.
- Snaydon. R. W. (ed.). 1987. Managed grasslands. Series: Ecosystems of the world. Amsterdam ; New York : Elsevier
- Taub, F.B. (ed.). 1984. Lakes and reservoirs. Series: Ecosystems of the world. Amsterdam ; New York : Elsevier.
- van der Maarel, Eddy (ed). 1993. Dry coastal ecosystems. Series: Ecosystems of the world. Amsterdam, The Netherlands ; New York : Elsevier.
- Whittaker, Robert J. 1998. Island biogeography : ecology, evolution, and conservation. Oxford; New York, Oxford University Press, 1998.

Ecotest

1. __________________ is the study of the geographical distribution of species.
a. Biodiversity
b. Biogeography
c. Environmental science
d. Biome

2. In terms of numbers of plants and animal species (biodiversity), the world's richest terrestrial biome is called:
a. desert
b. temperate grassland
c. tropical rainforest
d. tundra
e. taiga

3. The first application of biogeography was when A. R. Wallace subdivided the Earth's environments into six ___________.
a. packs
b. biomes
c. habitats
d. realms
e. taxa

4. The _______terrestrial biome has the following characteristics: cold climate; high latitude or high elevation; dominated by grasses, sedges, dwarf herbaceous species; treeless; year-round freeze potential.
a. tundra
b. taiga
c. rainforest
d. temperate forest
e. desert

5. The _______terrestrial biome has the following characteristics: moderate climate; deciduous trees are dominant; moderate tree species richness.
a. tundra
b. taiga
c. rainforest
d. temperate forest
e. desert

6. The _______terrestrial biome has the following characteristics: warm, moist or wet year round, large trees; highest tree species richness; numerous epiphytes, vines, and climbing plants; high diversity of all major life forms.
a. tundra
b. taiga

c. rainforest
d. temperate forest
e. desert

7. In the Florida Keys, the Australian pine has become a pest species. This non-native tree species was introduced decades ago and now is a dominant feature of the landscape, displacing native species. In Florida, the Australian pine is a(an) __________ species.
a. cosmopolitan
b. ubiquitous
c. exotic
d. endemic

8. The _______terrestrial biome has the following characteristics: extremely dry, warm or cool; plants are widely spaced and have specialized morphology and physiology to reduce moisture loss; many reptiles are well adapted to desert environment
a. tundra
b. taiga
c. temperate rainforest
d. deciduous forest
e. desert

9. The _______terrestrial biome has the following characteristics: cold winters, cool summers; dense stands of conifers (spruce, fir, and larch); low tree species richness.
a. tundra
b. taiga
c. temperate rainforest
d. deciduous forest
e. desert

10. Which of the follow islands would be expected to have the highest biodiversity?
a. a large island far from the mainland
b. a small island far from the mainland
c. a small island near the mainland
d. a large island near the mainland

11. Cosmopolitan species are found:
a. worldwide, wherever they are capable of living.
b. in the cosmos, or outer space.
c. everywhere, regardless of habitat.
d. in large cities.

12. The _______ aquatic biome has the following characteristics: saturated soils or shallow water; anaerobic, often organic soils; plants ranging from hardwood forests to grasses.
a. wetland
b. intertidal
c. freshwater
d. open ocean
e. hydrothermal vent

13. The _______ aquatic biome has the following characteristics: harsh environment due to wave action and alternating exposure and inundation by saline water; algae and mollusks attached on rocky substrates; on sandy substrates burrowing animals dominate.
a. wetland
b. intertidal
c. freshwater
d. open ocean
e. hydrothermal vent

14. The _______ aquatic biome has the following characteristics: hot, sulfur-rich water from tectonic vents stimulates growth of chemosynthetic bacteria that are the basis of food chain.
a. wetland
b. intertidal
c. freshwater
d. open ocean
e. hydrothermal vent

15. The _______ aquatic biome has the following characteristics: nutrient-poor, low productivity per unit area; pelagic organisms occur at low densities; global productivity is high due to vastness of surface area.
a. wetland
b. intertidal
c. freshwater
d. open ocean
e. hydrothermal vent

Chapter 9 - BIOLOGICAL PRODUCTIVITY AND ENERGY FLOW

The Big Picture

Biological productivity and energy flow are important quantities for ecologists to measure because they determine the rate at which biological resources such as trees are naturally replaced. For example, in the Case Study, two methods of forest harvest are discussed: (1) massive clearcutting was used in Michigan during 1860 - 1920, eventually causing nearly 99% of the original 7.7 million ha to become deforested permanently; and (2) the selective harvest on 14 % of a small 32.4 ha tract of woodland, which was maintained indefinitely in medieval England. The difference between the two methods was not only one of scale, but the degree to which the natural tree replacement rate was exceeded by the harvesting. An analogy can be drawn with a bank account, in which the harvest of tree biomass is equivalent to withdrawals, and the biological production is equivalent to increases due to interest accumulation on the biomass of trees present. In the Michigan situation, the account became overdrawn (withdrawals exceeded interest production). The result is a "stump forest" landscape that will not regrow into another forest. It is important to measure the rate of biological production so that we understand what sorts of biological limits will be placed on human harvests. In this chapter, the authors explain how ecologists estimate biological production and trace energy flow through an ecosystem. Other related topics to be discussed include: how much energy can be transferred between trophic levels; what the major physical laws of thermodynamics are and how they constrain energy flow in ecological systems; how ecosystems are one-way and open dynamic systems; how much energy is stored in biomass within the ecosystem; and how much energy is lost to randomness or heat in the system.

Frequently Asked Questions

What is energy?

- Simply stated, energy is the ability to do work or to move matter. It cannot be felt or observed, except as light or heat. Heat is a low quality (disordered) form of energy.
- Energy is a difficult concept to explain, yet everyone knows this word. In fact, most people expend a great deal of their income on energy in the form of electrical power, natural gas, or oil.

How is energy measured?

- Energy contained in organic matter can be measured by burning it and measuring the heat released. This is often done by placing a sample of known mass in a bomb calorimeter, a device that can be completely sealed and is insulated to prevent heat loss. A thermometer is placed inside (but it can be read from the outside) and the increase in temperature after the sample is "bombed" or burnt completely is measured. From this device, a series of measurement standards have been developed.

- Here are some standard units of energy:
  - **kilocalorie (or kcal)** = 1000 calories = the energy required to raise 1 kg of water 1.0 °C. (*Note: most food cartons and diet books incorrectly use the term "calorie" when they are in fact referring to kcal*).
  - **kilojoule (or kJ)** = 1000 Joules = 0.24 kcal. (The kJ is preferred over kcal as a scientific unit of energy).
- Here are some typical organic materials in an ecosystem and their energy content per gram:

Organic material	Energy content (kJ/g)
Fat	38
Muscle	25
Vegetation	21
Roots	19
Wood	17

- Thus, we can calculate the energy content of everything in an ecosystem.

What is biomass?

- Biomass is simply the mass of something that is or was alive. The mass is the quantity of matter in an object, or its weight. It is normally measured in grams (1 g = 0.035 oz.) or kilograms (1 kg = 1000 g = 2.2 lb.).
- Strictly speaking, mass and weight are different measurements. Weight is a function of gravity as well as mass; in space people are "weightless" but they still contain the same mass. On the surface of the Earth, mass and weight are identical numerical quantities.
- For ecological studies it is often easier to measure biomass when taking samples from an ecosystem. Later, biomass values can be converted to energy values using a table of conversion factors based on bomb calorimeter studies as described above.

What is photosynthesis?

- Photosynthesis is the process by which photoautotrophs make energy from sunlight. It can be written as a chemical equation:

$$6CO_2 + 6H_2O \text{ --------------> } C_6H_{12}O_6 + 6O_2$$

- This equation can be stated in words: "Six molecules of carbon dioxide combined with six molecules of water produces one molecule of glucose and six molecules of oxygen gas."
- Photosynthesis must take place under direct sunlight in the autotroph's tissues and chlorophyll or another photosynthetic pigment must be present for the reaction to occur. In this process, light energy is converted to chemical energy, which is stored in the bonds of the glucose molecule.

What is respiration?

- Cellular respiration is the process by which cells break down the glucose needed to provide energy for the cells. It is essentially the reverse of photosynthesis:

$$C_6H_{12}O_6 + 6O_2 \text{ --------------> } 6CO_2 + 6H_2O$$

- In this process, energy is released for the cell to do work.

What is biological production?

- Biological production is the amount of biomass in g or energy in kJ produced in an ecosystem per unit area and per unit time.
- There are two types of biological production: **primary production** (autotroph production) and **secondary production** (heterotroph production).

What is primary production?

- This is energy production by the autotrophs in an ecosystem. There are two kinds of primary production: **gross primary production** and **net primary production**.
- Gross primary production and net primary production are related in the following formula:

gross primary production - respiration = net primary production

or

gross primary production = respiration + net primary production

(These two equations are equivalent).

- Respiration is the energy used by cells in the autotrophic organism. The autotrophs produce more than enough energy for their cells' respiratory needs, so gross primary production (GPP) is normally greater than respiration (if it is less than respiratory energy needs, the autotroph will not grow and will eventually die).
- Thus, the net primary production (NPP) is the difference between the energy produced in GPP and the energy used in respiration. An analogy can be made with a business, in which the gross production is like the gross income for the business and the respiration is analogous to the fixed costs of conducting the business (rent, labor, taxes, etc.). The difference between gross income and the fixed costs for any period is the net income (or profit), which is analogous to net primary production in our ecosystem.

- In practice, ecologists measure GPP using the second equation above by calculating the net primary production, which is measured as a change in biomass per unit time, and adding that to a measurement of respiration:

$$\text{net primary production} = B_2 - B_1$$

where

B_1 is the autotroph biomass at time 1

B_2 is the autotroph biomass at time 2

- Respiration measurements can be obtained by conducting carbon dioxide or oxygen uptake studies.

What is secondary production?

- Secondary production is the amount of production per unit area and per unit time for heterotrophs. When heterotrophs consume autotrophs, the population and individual bodies of the heterotrophs will grow.
- The growth rate of heterotrophs is a good measure of net secondary production.
- Because heterotrophs respire as well as autotrophs, net secondary production and gross secondary production are related in a similar way as in the primary production equations:

$$\text{gross secondary production} = \text{net secondary production} + \text{respiration}$$

$$\text{net secondary production} = B_2 - B_1$$

where

B_1 is the biomass of heterotrophs at time 1

B_2 is the biomass of heterotrophs at time 2

What is carbon storage?

- Carbon storage is another unit for measuring biological production. Because carbon is included in molecules in the photosynthesis and respiration equations given above, and because of the fact that all organisms consume glucose to power their cells and make various other biomolecules from carbon in their bodies, carbon is a universal atom associated with life on Earth.
- Every organism stores carbon as it grows. It is thus a good measure of the photosynthesis and respiration rates, or the balance between them. Carbon can be easily determined because it remains behind after drying a sample in a furnace and weighing the remaining ashes. Often, production rates are given as g of carbon/m^2/year.

What is energy flow?

- Energy flow is the movement of energy through an ecosystem.
- Ecosystems are open systems, that is, they receive energy input from the sunlight, but that energy is eventually lost from the system. The energy is passed on to the autotrophs, to the heterotrophs, and eventually the energy is dissipated as heat out into space.

- Organisms (including dead and partially decomposed ones that are responsible for the oil, coal, and natural gas reserves on Earth) can store this energy for a while, but eventually it all flows out into space again.

What is thermodynamics?

- Thermodynamics is the study of the movement of energy and heat through systems, and the conversion of light, chemical, and mechanical energy into heat and vice versa.
- This science is responsible for most of the technology we call heating, refrigeration and air conditioning systems. It is one of the most fundamental of all sciences because it unifies physical, chemical, and biological principles.

What is the First Law of Thermodynamics?

- Energy can neither be created nor destroyed, but it may be converted from one form to another.
- For example, when the sun hits the sandy beach, it turns the sand warm. This is a thermodynamic conversion of light energy to heat energy. None of the sun's energy is destroyed, but some of the sun's rays are reflected back into space, some are converted to heat, some of the heat warms the air above the sand, etc.
- This law is also known as the **Law of Conservation of Energy**.

What is the Second Law of Thermodynamics?

- In any conversion of energy from one form to another (say, light energy to chemical energy), some amount of energy will be dissipated as heat.
- Thus no energy conversion is 100 % efficient. This is why there are no perpetual motion machines.
- If we took a 1000 kJ candy bar and burned it in a closed bomb calorimeter, there would be more heat in the chamber after burning (the internal temperature would rise), but there would still be 1000 kJ of energy in the system. Because the total energy is constant, and the amount of energy in the form of heat is greater than before, there must be less chemical energy left in the candy bar. The candy would have been burned away, heat produced, and the chemical bonds in the candy would be broken. The candy bar itself would have less mass as well, but the mass would be conserved in the entire system because of the next law of physics.

What is the Law of Conservation of Matter?

- In any physical or chemical change, matter is neither created nor destroyed, but it may be changed from one form to another.
- In our candy bar example above, the molecules of sugar (glucose) in the candy would be broken apart by the burning, but they would end up as CO_2 in the air in the chamber. In fact, for every one glucose molecule (which has six carbon atoms) we would find six carbon dioxide molecules in the air (each of which has one carbon

atom). So the carbon atoms would be conserved, and so would every other atom. Thus, no matter would be lost during this conversion of the candy bar into heat.

What is high quality energy?

- Light energy and chemical energy are high quality energy, because the energy is concentrated in a small space. A little bit of light or chemical energy can do a great deal of work. The molecules or particles that store these forms of energy are highly ordered and compact, thus it is high quality energy.

What is low quality energy?

- Heat is low quality energy. It can still be used to do work (think of a heater boiling water), but is much less compact and it rapidly dissipates. The molecules in which this kind of energy is stored (air and water molecules) are more randomly distributed than the molecules of glucose in our candy bar. This disordered state of the molecules and the dissipated energy requires a classification of low quality energy.

What is entropy?

- Entropy is a measure of the randomness and disorder of any system.
- Entropy will increase over time, unless energy is expended to reverse the disorder. Think of the laundry in your room. Unless you expend energy (do work) to wash it and fold it, it will become more disordered over time. In a similar way, all systems, including ecosystems, are becoming more disordered over time.
- Ecosystems only retain an order at all because of the energy input from sunlight and the work done by the autotrophs to make glucose. If the sun were to stop shining (as it is predicted to do in 5 billion years), then the ecosystems on Earth would become totally disordered and cease to function.
- The increase in randomness and disorder over time is called "an increase in entropy."

What is energy efficiency?

- Energy efficiency is the ratio of output to input, or the amount of useful work obtained from a given amount of energy. After every energy conversion, some energy is converted into low quality heat energy and the entropy of the system increases.

What is trophic level efficiency?

- It would be interesting to measure the efficiency of the transfer of energy between trophic levels. To do this, ecologists measure trophic level efficiency, which is the ratio of amount of biological production at one trophic level divided by the biological production at the trophic level right below it in a food chain. Usually this is expressed as a percentage of the lower trophic level. A good formula to use is as follows:

$$\text{\% Trophic level efficiency} = 100 \times \frac{[\text{Production at trophic level n}]}{[\text{Production at trophic level (n - 1)}]}$$

- For example, if the production of herbivorous rabbits is 1100 kJ/m^2/year, and the production of the grasses they feed upon is 12,000 kJ/m^2/year, then the % trophic level efficiency is:

$$\% \text{ Trophic level efficiency } = 100 \times \frac{\left[1100 \text{ kcal} / \text{m}^2 / \text{year}\right]}{\left[12{,}000 \text{ kcal} / \text{m}^2 / \text{year}\right]} = 9.1\ \%$$

- Most studies of trophic level efficiency indicate efficiencies of 10 % or lower. The efficiency of transfer of solar energy to green plants is 1-3 %. Wolves feeding on moose in Isle Royale National Park have a trophic level efficiency of 0.01 %. In managed agricultural ecosystems, the trophic level efficiency may approach or exceed 10 %.

What is growth efficiency?

- Sometimes ecologists wish to know "What is the efficiency of the transfer of energy *within* a trophic level?" To estimate this, they use a measure known as growth efficiency. There are two kinds: **gross** and **net growth efficiency**.
- **Gross growth efficiency** (also called the gross production efficiency) is the ratio of the growth of a consumer (in g) to the food consumption by that consumer (also in g).
- For instance, it takes an average of 7 lbs. of grains and human-edible plant material to produce 1 lb. of all livestock; this is a growth efficiency of approximately 14 %. For cattle, this growth efficiency is even lower: 6 % (16 lbs. of feed for 1 lb. of meat). For chickens and eggs, it is 33 % (3 lbs. of feed for every 1 lb. of eggs or meat). (Your textbook refers to these as trophic level efficiencies, but they are also referred to as growth efficiencies).
- **Net growth efficiency** (also called the net production efficiency) is the ratio of the growth of a consumer to the amount of food assimilated by the consumer (this is less than the amount consumed, because some food is never assimilated and is passed out as feces).

Which is greater, growth efficiencies or trophic level efficiencies?

- Growth efficiencies are typically greater than trophic level efficiencies (see Table 8.2). For example, whereas gross growth efficiencies are 8-27 % for terrestrial invertebrate herbivores (insects), net growth efficiencies are 20-40 %.

How does biological production and biomass change as energy flows up a food chain?

- The biological production and biomass at each higher trophic level decreases. This is due to several factors:
    - some of the energy is dissipated as heat and entropy increases at each energy transfer (the Second Law of Thermodynamics);
    - biomass is lost through respiration as organic carbon compounds like glucose are converted to carbon dioxide, which escapes into the atmosphere;
    - not all the available food at one trophic level is consumed by organisms at the next level, much of it instead going to the decomposers;

- not all the food consumed is assimilated by the consumer, some of it passing out as feces.

Should people eat lower on the food chain?

- Some environmentalists argue that humans (especially many overweight Americans) eat too high on the food chain and that by eating more fruits, vegetables and grains we could feed more people in the world than we do now.
- In order to prove this to yourself, consider that trophic level efficiencies are typically less than 10 %. This means that 90 % (and usually much more than this) of the food at each trophic level is not transferred to higher levels.
- If you eat nothing but tuna (a top carnivore in marine food webs at level 5), then you are consuming a lot of your energy at trophic level 6 or higher. If instead you consumed catfish (an omnivore at level 3), you would get the same amount of energy at trophic level 4 or higher. How much more energy is unavailable for other humans by eating the tuna? Consider the following data:

Assume: 10 % trophic level efficiency, and you eat 1,000 kJ of either type of fish:

Trophic level	Tuna (feeds at trophic level 5)	Catfish (feeds at trophic level 3)
6	You eat here	
5	1,000 kJ	
4	10,000 kJ	You eat here
3	100,000 kJ	1,000 kJ
2	1,000,000 kJ	10,000 kJ
1	10,000,000 kJ	100,000 kJ

- Thus, you have consumed 100 catfish dinners by consuming your one tuna dinner, because the primary production energy required to produce that tuna was 100 times greater than the primary production energy to produce the catfish (compare the energy levels at trophic level 1 in the table above).
- Thus, if you want to make more food available for others, you might choose catfish rather than tuna, because 99 more meals can be obtained with the same amount of primary production. This illustrates the concept of "eating lower on the food chain."
- However, there are other considerations, such a market price (tuna generally costs more than catfish), palatability (what tastes good to you), cultural factors (such as whether catfish or tuna has been a traditional food item in your culture), and availability (maybe there isn't any catfish on the menu in the restaurant where you're dining). You may wish to lower your intake of toxic chemicals such as mercury, which tends to biomagnify (see Chapter 14), and will be in a higher concentration in the tuna. In addition, eating less tuna does not automatically result in more catfish biomass -- the primary production of the ocean where the tuna feeds is not easily moved into a catfish farm (but some of it is, because catfish are being fed soybeans plus menhaden oil, from another kind of ocean fish. This is why catfish are considered to be at trophic level 3).

- What should you do? The choice is up to you, but at least consider the food chain level at which you eat. (To determine the trophic level at which you feed most of the time based on your own diet, please see Ecology in Your Backyard in Chapter 6).

Ecology In Your Backyard

- **Measure primary production in your backyard.** This procedure involves "destructive sampling", as ecologists refer to it, but you probably call it "mowing the grass"! In order to estimate net primary production on your lawn, we will have to measure the change in biomass over time. A good period over which to measure growth is two weeks. Cordon off a 1.0 m^2 area of your lawn (the area can be any size, but you should measure the area and then divide your biomass estimates by the area you select). Cut down the grass as close to the earth as possible (don't worry, tell your landlord or parents that it will grow back). Use scissors, hand-held garden clippers, or if you are careful to save all the grass blades, a power mower. Weigh the grass blades you have cut on a postal balance or some other accurate balance (a bathroom scale can work well here if the area is large and your clippings weigh a lot). This represents the standing (above-ground) biomass on that plot of lawn. (We will ignore below-ground biomass and productivity for this experiment, because you would have to dig up all the roots.) Now wait two weeks for the grass to grow back and cut it again close to the ground to the same approximate level as before (you may choose a longer or shorter time interval depending on the rate of growth on your lawn, fertilizers used, soil nutrients, sunlight, water, etc.). Weigh the clippings again. This represents the net primary production in $g/m^2/2$ weeks for your lawn (if you weighed in ounces, multiply by 28.35 to obtain the weight in g; if you weighed it in pounds, multiply by 453.6 to get g). You can multiply this value by 21 kJ/g to get the total energy in the grass clippings. You could also multiply these productivity values by 26 to get an annual production rate (there are 26 two-week periods in a year), but this would assume that growth you measured was constant across all seasons, which we suspect is not true. A better way would be to repeat the measurement on a new plot of lawn every two weeks for an entire year. You would find that the primary production slows down in the winter and is the greatest in the summer. You could also use this measure of net primary production to test the impact of various light levels, watering regimes, fertilizers, herbicides, pesticides, etc.

- **Measure the growth efficiency of your dog, cat, or other pet**. This is a harmless procedure which involves feeding your pet (you do that anyway!), and keeping careful notes about the pet's changes in weight or biomass. You will need a balance or bathroom scale to weigh the food and the pet. A food scale used by people who diet or a postal scale can be used for weighing the food. You do not have to weigh every meal eaten by your pet, but try to feed standard servings (1/2 can of food, one scoop or bowl full of kibble, etc.), and weigh that once the first time you feed (you may wish to improve your accuracy by weighing several standard servings and taking the average). Then later during the growth study you can simply count servings and multiply by this "average" serving weight. Try and weigh or estimate the weight of any uneaten food left behind. Let your animal feed on as much as it wants, i.e., don't restrict its diet other than you normally would. Your veterinarian may be able to help you get an accurate weight of your pet at the beginning and end of the study, so you can calibrate your home scale. This works best with puppies and kittens or other animals that are in their major period of growth; adult animals will show little growth, and so your growth efficiencies will be very small or zero (if they are significantly less than zero, i.e., -10 %, your pet might be sick and you should take it in to see a veterinarian). Fill out the following table (be sure to use the same biomass units throughout or convert to common units. If your animal is small, then use oz. or g; if it is large, then use lb. or kg):

Week	Animal weight at start of week (B_1)	Animal weight at end of week (B_2)	Growth during week (B_2 - B_1)	Food consumed during week (C)	% Growth efficiency = [100 * (B_2 - B_1)/C]
1					
2					
3					
4					

Links In The Library

- Cousins, S. 1985. Ecologists build pyramids again. New Scientist. 4 July 1985. 106
- Lindeman, R. L. 1942. The trophic-dynamic aspect of ecology. Ecology 23: 399-419. A classic paper that began the movement to study food webs and trophic interactions.
- Pimm, S.L., J. H. Lawton and J. E. Cohen. 1991. Food web patterns and their consequences. Nature 25 April 1991, 350 (6320): 669.
- Polis, G. A. and K. O. Winemiller. 1996. Food Webs: Integration of Patterns and Dynamics. Chapman & Hall, New York. 472 pp.

Ecotest

1. The fact that energy losses from one trophic level to the next are about 90 percent is explained by the:
a. second law of thermodynamics.
b. law of conservation of matter.
c. first law of thermodynamics.
d. None of these choices are correct.

2. Of the following types of heterotrophs, which has the lowest net growth efficiency, according to the textbook?
a. microorganisms
b. invertebrates
c. vertebrates
d. All of these have similar net growth efficiencies.

3. Calculate the trophic level efficiency of energy transfer between a population of rabbits that has a secondary production (or growth rate) of 2000 kJ/m^2/year and a population of wolves that has a tertiary production of 10 kJ/m^2/year.
a. 5 %
b. 10 %
c. 0.01 %
d. 0.5 %
e. 0.005 %

4. If the Gross Primary Production (GPP) for a marine ecosystem is 100,000 kcal/m^2/year and the respiration for this ecosystem is 90,000 kcal/m^2/year, what is the Net Primary Production (NPP) for this ecosystem?
a. 10,000 kcal/m^2/year
b. 100 kcal/m^2/year
c. 0.9 kcal/m^2/year
d. 0.1 kcal/m^2/year

5. Calculate the gross growth efficiency for snook population (snook are carnivorous fish that eat shrimp) that consumes 5000 kJ/m^2/year of shrimp and grows 250 kJ/m^2/year.
a. 25 %
b. 10 %
c. 5 %
d. 1 %

6. Why is biomass at the second trophic level in most ecosystems not as great at as biomass at the first trophic level?

a. All of these are correct
b. Not all plant material is eaten by herbivores.
c. Not all of the biomass eaten by herbivores is digested.
d. Some of the biomass is used to make cellular energy and ultimately is respired.

7. If you have a choice between consuming a beef hamburger and a tunafish sandwich, and you want to eat "lower on the food chain," which would you choose?
a. hamburger
b. tunafish sandwich
c. Neither one, because they are at the same trophic level.
d. None of these are correct

8. The amount of randomness and disorder increases as ___________________ increases in a system.
a. energy
b. biomass
c. entropy
d. biological production

9. Which of the following contains the lowest quality of energy?
a. gasoline
b. sugar
c. a car battery
d. a mug of hot coffee, with no sugar

10. You have a group of rabbits and their total weight at the start of the year is 10,000 g. After the year ends, you weigh them again and find that they now weigh 15,000 g. What is the secondary production for the group during the year?
a. 66.7 %
b. 33.3 %
c. 50.0 %
d. 5,000 g/year

11. Complete the equation: $6\ CO_2 +$ ______ —— $C_6H_{12}O_6 + 6\ O_2$
a. $6\ OH^-$
b. $6\ CH_4$
c. $6\ H_2O$
d. $12\ H_2O$
e. O_3

12. The growth of all the herbivores in an ecosystem is an approximate measure of _______________.
a. net primary production
b. carbon storage
c. net secondary production

d. gross primary production
e. all of the above

13. The study of the movement of energy and heat through systems is called:
a. hydrology
b. fluid mechanics
c. conservation biology
d. thermodynamics
14. The fact that there is no such thing as a "perpetual-motion machine" is explained by:
a. the first law of thermodynamics.
b. the second law of thermodynamics.
c. the third law of thermodynamics.
d. the law of conservation of matter.

15. Which of the following choices shows a decrease in the amount of entropy in the system?
a. A bag of marbles breaks open, and the contents spill out onto the floor.
b. An electric fence loses power, allowing the cattle held within to escape.
c. You forget to cut your grass for a month, and return to find several new species of weed present.
d. You return all of the books you have been reading to your bookshelf.

Chapter 10 – ECOLOGICAL RESTORATION

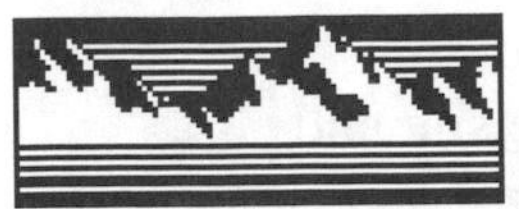

The Big Picture

When viewed from an airplane, much of the landscape of the United States is a patchwork. There is a natural patchiness to the landscape resulting from differences in soil characteristics, topography, microclimate, fire, and biotic factors. Superimposed on this is the modern landscape that has been shaped by humans. For much of the United States, a mosaic of fields, forests, lakes, roads, towns, and cities dominates the landscape. To an ecologist, much of the patchwork represents communities in different stages of succession. For example, from a successional perspective, a cornfield is a pioneer community, this "grass" stage that develops soon after disturbance (plowing) and, as long as the land is kept in agricultural production by annual plowing and planting, such pioneer communities will persist. If the land were no longer cultivated the site would eventually revert to forest. Similarly, the rice paddies of Southeast Asia, the sugar cane fields of Cuba, slash and burn plots in the Amazon Basin, pastures of Ireland, and pine plantations of southeastern United States are all communities in early stages of succession. Throughout large sectors of the world, humans have manipulated succession to provide necessary food, fiber, and natural resources. In these areas, only a fraction of the landscape has not been altered. In the eastern United States, less than one percent of the old-growth, virgin forests still exists. While much of the eastern landscape is forested today, these forests are still recovering from being logged as much as three or four times over. Actually there is more eastern land in forest cover today than at the beginning of the century. A century ago Vermont was 30 percent forest and 70 percent cleared (mostly pasture); today these numbers are reversed. Many other ecosystem types have disappeared from the landscape. The tall-grass prairie consists of a few isolated remnants, the grasslands of central California have been converted to cropland or urbanized, coastal development has obliterated the barrier island communities of the East and Gulf coasts, and wetlands of all types have been reduced by at least fifty percent. There have been attempts at ecosystem restoration. For moral, aesthetic, economic, legal, and ecological reasons private and public land managers are replanting and restocking native plants and animals in their former positions in the landscape. If restoration is to be successful, it is imperative to have a sound understanding of ecological succession.

Frequently Asked Questions

What is ecological succession?

- Ecological succession is the natural process of establishing or reestablishing an ecosystem.
- Succession occurs as a series of plants, animals, and microbes colonize a site over time.
- Succession may occur on sites that were previously unoccupied or on sites at which the exiting biotic community was removed or altered by disturbance.

- **Primary succession** occurs on sites that have not been previously occupied. Such sites exist on newly formed soils, such as exposed sandbars and volcanic ash. The colonization of a sunken ship by coral reef organisms is an example of primary succession.
- **Secondary succession** occurs when a biotic community has been disturbed, then becomes reestablished. Remnants of the previous biotic community still exist at the site and contribute to recolonization. Secondary succession is much more common than primary succession; much of the landscape is in a stage of secondary succession. One common example of secondary succession is the reestablishment of forest vegetation after an area has been logged. When a forest is logged, only the merchantable timber is removed; undesirable tree species, shrubs, saplings, and seeds remain and constitute the colonizer pool.

What is the role of disturbance in succession?
- Ecological disturbance is an event that disrupts ecosystem, community, or population structure.
- Disturbance may be initiated by natural events or human activities.
- Common agents of natural disturbance include fires, hurricanes, tornadoes, landslides, floods, droughts, and biogenic disturbance (caused by animals such as prairie dogs, elephants, and outbreaks of insect pests).
- Human-initiated or anthropogenic disturbance dominates the landscape in much of the world. Anthropogenic disturbance includes such activities as the conversion of natural communities to farm field and pasture, logging and deforestation, surface mining, urbanization of natural landscapes, and warfare.
- Disturbance creates ecological opportunities for individuals to become established that would otherwise be excluded from the community.
- Species diversity can be strongly influenced by disturbance. Highest species diversity is often found in ecosystems where disturbance is neither too rare nor too frequent; similarly neither too benign nor too severe. This is known as the *intermediate disturbance hypothesis* for species diversity.

What are the stages of succession?
- Biotic communities change over time. The complete sequence of change is called a *sere*.
- Seres are made up of recognizable units called *seral stages*, although the transition from one seral stage to the next is gradual and intergraded.
- Species that appear in early successional stages are called pioneer species. **Pioneer species** tend to be fast growing, short-lived, and capable of rapid and wide dispersal. The grasses and herbaceous plant species that colonize a farm field soon after its abandonment would be considered pioneer species.
- Late successional species tend to be persistent, longer-lived species.

What patterns develop as seral stage changes?
- Ecologists recognize at least four patterns or models of succession.
- **Facilitation**: Early seral stages modify microclimatic and soil conditions that facilitate colonization by later seral stage species. Early successional species so modify the site that

the conditions that allowed them to colonize initially are no longer present, thus their propagules (progeny) become excluded from the site.

- **Interference**: Early seral stage species colonize a site and inhibit colonization by later stage species. This is contradictory to the facilitation model. Some early species exude toxins into the soil to prevent other species from occupying the site; this is called *allelopathy*. Even though these early stage species may persist for an extended time, their shorter longevity will allow late stage species to eventually colonize.
- **Life History Differences**: The occupation of a site is a consequence of the life histories of the colonizing species. Means of dispersal to a site and the successful establishment of propagules differs among species.
- **Chronic patchiness**: In harsh environments or often disturbed environments, early successional species are often favored and late successional species are not able to persist. A familiar example would be the maintenance of grassy lawns (early seral stage species) by frequent mowings (disturbance). Abandoned lawns are soon colonized by shrub and tree species (late seral stage species).

How does succession affect biogeochemical cycling?

- In addition to changes in species composition that occur during succession, other attributes of ecosystems change, including biogeochemical cycling.
- In the early seral stages of terrestrial succession, plant and animal biomass is low and productivity (growth rate) is high. Inorganic nutrients (e.g., N, P, K, Ca) are incorporated into biomass rather than being leached from the ecosystem. As the community matures, organic matter and nutrients are returned to the soil and again become available for plant uptake; this is called *internal nutrient cycling*. Dead organic matter and live roots retard soil erosion and thus reduce the associated nutrient loss.
- Biomass and soil are nutrient reservoirs or *nutrient sinks*.
- Maximum nutrient storage is achieved during the middle stages of succession while both productivity and soil nutrient storage are high. As the ecosystem ages further and productivity declines, there is a net loss of nutrients because productivity does not keep pace with nutrient losses due to soil erosion.
- The nutrient retention of the soil is dependent upon the amount of live and dead organic matter present, and soil texture (the size of soil particles). Nutrients are easily leached from large grained soil particles (sand) but are more readily retained in fine grained soil (clay or silt). *Loam* is an even mixture of clay, silt, sand, and well decomposed organic matter that maximizes soil nutrient retention and accessibility, and permeability thus is quite favorable for plant growth and root development.

How does disturbance affect biogeochemical cycling?

- Disturbance plays an important role in biogeochemical cycling.
- Fire releases nutrients stored in biomass. As ash, these nutrients may be removed from the ecosystem by wind and water; but nutrient-containing ash that remains on-site becomes available for new plant growth. A pulse of available nutrients causes a pulse in plant growth soon after fire.

- Disturbance events such as floods and storm waves may relocate massive amounts of plant debris and other organic materials (and nutrients) downstream or along high tide or flood lines (also called wrack lines).
- Storms that topple trees create canopy gaps. Below canopy gaps, increased light on the forest floor stimulates plant growth and nutrient uptake. The "tip-up" mounds created by the roots of fallen trees transfer nutrients from soil depth to surface.
- Anthropogenic disturbances that disrupt plant cover or soil structure cause nutrients to be lost from ecosystems. To produce crops, fields are cultivated, which results in increased erosion and nutrient loss; additionally, the harvest of plants or plant parts removes nutrients from the system.

What is a climax community?

- The theoretical end-point of a succession sere is the climax community. At this stage, the community is expected to be self-replicating and persist indefinitely. Individual plants and animals will die but are replaced by species within the same community, thus community composition remains at equilibrium.
- In addition, the classic theories of succession and climax community proposed that at climax, the community would have achieved maximum organic content, maximum chemical (nutrient) storage, and maximum biological diversity. Certainly, many old-growth communities exhibit these characteristics.
- But, ecologists now realize that the classic concept of climax community is not valid and that many of the old-growth communities formerly considered climax communities are not the end-points of successional seres.
- Maximum organic content, nutrient storage, and biological diversity are found in intermediate stages of succession, not at the end-point of a sere.
- The time scale of successional change is difficult to comprehend in human life spans. Successional transitions early in seres may be readily apparent within a few years or decades, however, communities that appear late in the sere may persist for many human generations and changes are not as evident.
- Climates are not stable over long periods of time. Consequently, communities respond as climate changes.

What are some examples of succession?

- **Succession in the Ocean** Succession in the ocean occurs at sites that have a relatively stable substrate such as on rock outcrops. A good example of primary succession in the ocean is the succession of coral reef species on ships that were sunk in the South Pacific during World War II (e.g. Truk Lagoon).
- **Aquatic or Wetland Succession** Lakes and ponds are temporary features of the landscape, although large and deep lakes, such as Lake Baikal in Russia, have persisted for millions of years. Lakes and ponds are sediment and nutrient sinks, thus they accumulate organic and inorganic materials on the lake bottom, becoming shallower and eventually filled. Terrestrial plants may eventually colonize sites that were once open water. The **bogs** of the Great Lakes states and other northern regions were so formed.

- **Oligotrophic** is a term used to describe the nutrient status of geologically young, nutrient-poor lakes and ponds. As sediments and nutrients accumulate and aquatic plant growth increases, the waters become **eutrophic**. This process is called **eutrophication**.
- Human activities that cause soil erosion or nutrient input may accelerate the processes of eutrophication and aquatic succession.
- **Sand Dune Succession** Wind and water cause the shifting sands of dunes to be geologically unstable. Pioneer plant species of sand dune ecosystems must anchor the sand with extensive root systems. Once stabilized, other plant species may colonize and begin to modify site conditions, facilitating the establishment of late successional species. In the absence of subsequent disturbance forest communities dominate the site. Sand dunes are usually located along lake or oceanic shorelines thus are exposed to erosive forces of winds, tides, and waves. Dunes nearest the shoreline are the most frequently disturbed and receive the most severe storm impacts. These fore dunes are maintained in a state of early succession. Further away from the shoreline, more stable communities develop. Shorelines change, however, fore dunes can be breached by storm waves, inlets can migrate, and the more stable, interior dunes can be disturbed.

What can be done to restore ecosystems?

- Ecosystems are unimaginably complex; to assume that an ecosystem can be completely restored by human efforts and technology is unrealistic and presumptuous. Nonetheless, attempts at ecosystem restoration are justifiable. In some circumstances, land managers may be legally required to restore ecosystems or managers wish to restore ecosystems for aesthetic, ecological, economic or moral reasons.
- Ecosystem restoration requires a sound understanding of ecology in order to determine what has been altered, why and how the system has been altered, and what steps must be taken to achieve the targeted restoration condition. The landscape is cluttered with species that have been introduced in an attempt to improve environmental quality but without understanding the ecological risks. Purple loosestrife, kudzu, mongoose, and European starlings are but a few species that have been introduced with good intentions but undesirable results.
- It is important to set realistic goals for restoration. For example, the reestablishment of a fire-dependent grassland ecosystem in a residential land use area is unrealistic.
- Ecosystem restoration must take into account social and economic conditions of the surrounding community. For example, the restoration of a single species, the gray wolf, in the Northern Rocky Mountains has precipitated considerable debate within the community.
- Most terrestrial ecosystem restoration entails planting appropriate species of vegetation in a manner that accelerates the successional process but that is ecologically sound, or creating the appropriate habitat conditions that will permit natural colonization of the site. The restoration of animal populations is primarily dependent upon natural colonization, but animals may be introduced if the appropriate habitat conditions have been established.
- The process of ecosystem restoration requires a thorough knowledge of site conditions (natural and land use history, soils, hydrology, species interaction) and the desired target condition. From this information a restoration plan must be developed.
- The restoration plan must be implemented in the proper sequence and during the appropriate season. For example, soil preparation would have to be accomplished before planting

begins, and planting must be started and completed within the appropriate seasonal time frame (usually winter) to maximize seed or sapling survival. Remedial plantings are often necessary to replace plants that have died.

- A strategy for weed and pest control is usually necessary. This may require the use of herbicides, pest or predator deterrents, cultivation, mechanical removal, biological controls, or a combination.
- If fire is to be used as a management tool, a burning regime must be implemented.
- In order to gauge restoration success, a monitoring program must be implemented.
- Constructed, restored or enhanced ecosystems have been used to mitigate (offset) environmental degradation resulting from development and land alteration; wetland mitigation is now a frequently employed strategy that attempts to compensate for wetland degradation.
- Long term protection of the site may be achieved through local, state, and federal incentive programs that provide tax breaks, cost share opportunities, and direct payments.
- Ecosystem restoration may be undertaken by individual landowners, environmental consulting contractors, environmental groups, and government agencies.

What are biomes?

- **Biomes** are major ecological divisions of the terrestrial environment (See Chapter 7). Biomes (Clements and Shelford) are but one of several classification schemes that have been developed for the Earth's biota; others include biogeographic realms (Wallace), life zones (Holdridge), and ecoregions (Crowley and Bailey).
- Biomes classification is based primarily on similarities in plant growth form. Growth form is in response to climatic factors, principally temperature (along a latitudinal gradient) and moisture (Figure 8.6). Although species differ within biomes, growth forms are consistent. For example, temperate grasslands occur in North America and central Asia; plant species are different between these regions but the dominant growth forms (grasses, sedges, and herbaceous plant species) are common to both.
- Biomes are discussed in this chapter because they represent the most stable seral stage for a given region and climate. In the absence of major disturbance by humans or natural factors, the biomes listed below would persist in the landscape.
- Most ecologists recognize at least nine major biomes; your text lists ten terrestrial biomes as well as major aquatic and marine systems (See review chart on next page).

Ecology In Your Backyard

- This exercise can be used to track primary succession. Fill greenhouse trays with sterile potting soil. Potting soil should be heated in an autoclave beforehand to kill any existing seeds. Place replicate trays in a natural area that is protected from human or animal disturbance. Make a data sheet to record biweekly counts throughout the growing season of seedlings that germinate in the trays. You will have to identify species of seedlings and count the number of individuals for each species. Seedling identification can be very difficult; if you cannot identify the species, make a sketch of each unidentified type of seedling and give it a name (i.e., Species A, Species B, Species C, etc.).

- Average your biweekly data, and calculate the Shannon Diversity Index and evenness values (See *In Your Back Yard* in Chapter 7). Record these values along with species richness. At the end of the growing season, generate bar graphs for each of these parameters.
- If you would like to participate in forest or prairie restoration projects or in an urban forestry project, there are several conservation groups you may contact. *The National Arbor Day Foundation* and *The American Forest, Global Releaf* programs are but two organizations specializing in ecosystem restoration. You may wish to contact your city parks or horticulture department of county extension office for additional information about on-going projects, volunteerism, and planting techniques.

Links In The Library

- National Research Council. 1992. *Restoration of Aquatic Ecosystems*. National Academy Press, Washington, D.C. 552 pp.
- Society for Ecological Restoration. 1982+ . A twice yearly publication by the University of Wisconsin Press for the Society for Ecological Restoration.
- Thayer, G.W. (ed.). 1992. *Restoring the Nation's Marine Environment*. Maryland Sea Grant College, College Park, Maryland. 716 pp.

Ecotest

1. When a forest is logged, remnants of the community (i.e., seeds and saplings) are left on site and contribute to site recolonization. This is an example of ________________ succession.
a. residual
b. cyclic
c. primary
d. secondary

2. The entire successional transition from an abandoned field to a forest is called a _____.
a. pattern
b. sere
c. series
d. procession

3. An event that disrupts ecosystem, community, or population structure is called a (an) ______ event.
a. interference
b. disturbance
c. non-sequential
d. inconsistent

4. Which statement about biogeochemical cycling during succession is false?
a. Maximum nutrient storage in soil occurs during the late or climax community stage.
b. Soil texture is a determining factor in nutrient storage.
c. In early seral stages, nutrients are rapidly incorporated into plant biomass, reducing the amount of nutrients lost from the ecosystem via leaching.
d. As communities mature from early to intermediate seral stages, internal nutrient cycling increases.

5. Lakes that are of poor nutrient status (i.e., low nutrient concentrations) are called ________ lakes.
a. eutrophic
b. oligotrophic
c. microtrophic
d. autotrophic

6. Short-lived species that are capable of rapid and wide dispersal and are usually the first species to colonize an unoccupied site are called _____ species.
a. phalanx
b. competitive
c. facultative
d. pioneer

7. In the _____ model of succession, early species modify the microclimate, making it more favorable for later colonizing species.
a. competition
b. tolerance
c. facilitation
d. mediation

8. If a site is frequently disturbed so that only early successional species can survive, this successional pattern is called:
a. patch dynamics.
b. chronic patchiness.
c. induced patchiness.
d. biogenic patchiness.

9. The ______ biome is characterized by cold winters, cool summers, dense stands of conifers (spruce, fir, and larch) and low tree species richness.
a. temperate forest
b. taiga
c. temperate woodland

d. tundra

10. The ______ marine environment is characterized by strong wave action and alternating exposure and inundation by saline water; algae and mollusks are attached on rocky substrates; on sandy substrates burrowing animals dominate.
a. intertidal
b. open ocean
c. benthic
d. upwelling

11. All of the following are temporary (appear and disappear in time scales < 10,000 years) features of the landscape except:
a. mountains.
b. sand dunes.
c. lakes and ponds.
d. old farm fields.

12. Maximum organic content, maximum biological diversity, and maximum nutrient storage occur in which of the following successional stages?
a. Early successional stage
b. Intermediate stage
c. Climax stage
d. None of these are correct

13. Ecosystem compartments that remove nutrients from cycling are called:
a. nutrient leaks
b. nutrient gaps
c. nutrient sinks
d. nutrient drains
e. nutrient pockets

14. In the classical view, a climax community is expected to be:
a. self-replicating.
b. the endpoint of succession, where community composition is at an equilibrium.
c. around indefinitely.
d. All of these are correct.

15. The natural process of establishing or re-establishing an ecosystem is called:
a. environmental degradation.
b. ecological succession.
c. ecological extinction.
d. restoration

Chapter 11 - WORLD FOOD SUPPLY

The Big Picture

The world's food supply is one potential limiting factor to human population growth. Will the Earth's human population exceed the food production of the planet? We already know that per capita food production is dropping in some parts of the world and for the world as a whole. This means that there is less food per person than there was just a decade ago. Some countries have already experienced famine (China, Bangladesh, Ethiopia, Somalia, and North Korea) and more will in the future. Some of these famines and food shortages are caused in part by social and political upheaval, and if these problems could be resolved, food production could increase. However, a common problem remains that food is unevenly distributed worldwide. There are nations with food shortages and nations with food surpluses, but no economical and convenient way to distribute the food rapidly between these nations when a shortage develops. In this chapter, you will examine world food production from an ecological perspective. You should try to imagine that you are a visitor from another planet, an ecologist, and that you are attempting to describe to people on your planet what you observed about the food supply in relation to the human population growth while you were on Earth. In this chapter, the authors describe how agro-ecosystems differ from other natural ecosystems, how much land is devoted to agriculture, how water availability can limit food production, how much advances in modern agriculture can improve food production, and how the growth of the human population causes deterioration of the very agro-ecosystems that support that population.

Frequently Asked Questions

What are the major foods that humans depend upon for sustenance?

- Most human food comes from land-based agriculture and livestock production, although about 5 % comes from fisheries and aquaculture.
- Of the 250,000 species of plants on Earth, about 300 are agricultural crops, but only 100 are grown for food on a large scale.
- Only 20 species of plants comprise the bulk of human food: wheat, rice, corn, potatoes, barley, sweet potatoes, cassavas, soybeans, oats, sorghum, millet, sugarcane, sugar beets, rye, peanuts, field beans, chick-peas, pigeon peas, bananas, and coconuts (listed in order of importance).
- Humans also eat livestock and they feed their livestock alfalfa hay, sorghum, other grasses, and fish meal.

How much food is produced per person per year?

- Humans currently produce 1,780,000,000 metric tons of grain per year.
- There are approximately 5,700,000,000 humans on Earth. This works out to 0.31228 metric tons/person/year or (1 metric ton = 1000 kg) 312.28 kg/person/year (or 688.5 lbs/person/year).

How much food is needed per person per year?

- In order to maintain a minimal level of activity, people need 200 kg/year; this is the borderline between undernourishment and healthy levels of food intake.
- Many people need more than this level to be active and perform work every day.
- Thus, on a worldwide average, there is more food produced each year (312 kg) than is needed per person (200 kg).

If more food is produced than is needed per person each year, why is there world hunger?

- Food is unevenly distributed worldwide. Some countries have a food surplus (like the U.S.), and others lack enough food or the infrastructure to distribute it.
- All countries except the U.S., Canada, and Australia are net food importers. Many developing countries grow cash crops rather than food crops to gain money to purchase food; thus, they are not self-sufficient when it comes to food production.
- However, many countries and people within those countries do not have sufficient funds to purchase food on the world market. In addition, food distribution systems (boats, trains, trucks, and stores) do not approach the levels that most Americans experience.
- Therefore, an unequal distribution of food results in famine and hunger, even though there is theoretically enough food for all.

What have been the recent trends in food production?

- The world's grain supply can be measured in terms of the number of days of grain supply on-hand.
- Since 1960, the number has fluctuated, reaching a low of a 56-day supply in 1973 to a high of a 104-day supply in 1987.
- The 1995 estimate was a 62-day supply.
- By this measure, the supply of grain has been declining.

Has food production exceeded human population growth?

- It appears that Malthus was correct in 1798 when he predicted that there would be a point when population growth would exceed food production.
- The per capita food production rate is currently declining, after nearly four decades of steady increases in this measure of food production. (See the Environmental Issue, "Was Malthus Right?" in Chapter 5)

How many kilocalories are required per person per day?

- About 2600 kilocalories per day are required for a minimum of activity and for a person to perform work.

- A more reasonable estimate is 3000 kcal/d for the average person, however this is a function of body size. In some climates, more energy may be required to make up for heat loss (Eskimos may need 7000 kcal/d).

What is undernourishment?

- Undernourishment means that insufficient energy is taken in each day to offset metabolic uses. Everyone has a baseline metabolic rate and cellular respiration which is always ongoing. This causes people to use energy even when sleeping.
- If people consume fewer than 2600 kilocalories per day, the baseline metabolic rate, then they are considered undernourished. Undernourished people have little ability to move or work. If they do not receive this baseline metabolic rate, eventually they will die.
- This is the condition that causes death during famines.

What is malnourishment?

- People that are malnourished, in contrast, may have sufficient energy intake in terms of kilocalories, but they are missing specific dietary nutrients, such as a lack of vitamins, proteins or **essential amino acids**.
- Malnourishment leads to chronic medical conditions such as **marasmus** or **kwashiorkor,** which do not kill people outright, but make them very unproductive individuals in society.

What is marasmus?

- Marasmus (from the Greek marasmos, meaning "wasting away") is a progressive emaciation of the body caused by a lack of proteins and calories.
- Symptoms include a pronounced slowing of growth and extreme atrophy (wasting away) of muscles.
- Occurs most frequently in children.

What is kwashiorkor?

- Kwashiorkor is a native word in Ghana, meaning "displaced child".
- In this condition, people simply do not get enough protein in their diet. They do receive sufficient calories, however. Because of the lack of protein, they develop neurological problems and learning disabilities.
- Symptoms include edema (swelling), especially near the abdomen; stunted growth; brittle, dry reddish hair; and apathy.
- Occurs most frequently in children.

What is an agro-ecosystem?

- This is an ecosystem that has been manipulated by humans to create special conditions in which the crops they prefer to eat can be grown in great abundances.
- These systems are very different than natural ecosystems and require great inputs of energy to maintain in this altered state.

How do agro-ecosystems differ from natural ecosystems?

- Agro-ecosystems have the following characteristics that distinguish them from other ecosystems:
- **Early successional stages** - The crops are planted in fields that have been plowed, which creates a disturbance event. The crops that are planted do well in this early successional state, but any colonizer plants that could compete with them are removed (humans call these weeds).
- **Monocultures** - This means that all the crops in a field are of one species or variety. Genetic and species diversity are low in agro-ecosystems. Soils become depleted of species-specific nutrients if one variety is grown in the same place repeatedly. Crop rotation and artificial fertilizers are used to restore soil fertility.
- **Low habitat complexity** - Agro-ecosystems are planted in rows, which facilitates harvest, but can also facilitate disease and pest infestations. In natural ecosystems, there is complexity in the distribution of the plant species, so that species-specific infestations cannot spread as easily.
- **Simple food webs and food chains** - In agro-ecosystems, the food web is early successional low diversity, so there are few top predators. Herbivorous insect species can easily become abundant without the top predators, and this leads to pest outbreaks. In natural ecosystems, there are many species in a web and the predators can hold the prey in check.
- **Frequent soil disturbance** - Plowing the soil represents a regular soil disturbance unlike anything in a natural ecosystem. Plowing increases soil erosion and nutrient losses.

What is crop rotation?
- Crops are planted in the same field but in a designed sequence over many years, so that the soil nutrients can be replaced or recover.
- For example, rather than planting corn in the same field year after year, if the fields are planted with soybeans in alternate years, the soil nitrate levels will be rebuilt. Soybeans are legumes that fix their own nitrogen.

What is traditional agriculture?
- Also called swidden, milpa, fang, or bush fallow agriculture.
- This is planting a diversity of crops in a cleared forest patch. No plowing, no rows, no pesticides, fertilizers, etc.
- This approach is called polyculture rather than monoculture. Something will always be fruiting and ready to harvest.
- This is sustainable agriculture at low human population densities.

What is a limiting factor?
- When soils lack a specific nutrient required by a plant, so that the plant cannot grow without it, that nutrient is called a limiting factor.
- These nutrients could be **macronutrients** such as sulfur, phosphorous, magnesium, calcium, potassium, nitrogen, oxygen, carbon, and hydrogen.
- They could also be one of the **micronutrients**, or trace elements, which are typically metals like molybdenum, copper, zinc, manganese, iron, boron, and chlorine. This is true in Australia, which has trace-metal deficient soils.

- It is also possible for there to be too much of each of these nutrients, and if this occurs, the plant will not grow either.

What is Liebig's law of the minimum?
- Justus von Liebig was a farmer and understood about limiting factors.
- He noticed that the abundance and distribution of a species of crop is limited by the single soil nutrient factor that is in shortest supply.
- His law has been extended beyond soil nutrients to many other factors and species, and is the basis for the science of ecophysiology.

How can agricultural production be increased?
- There are several ways in which agricultural production may be increased by humans:
  - **Increase the farmed land area** - We can attempt to farm additional land, but most farm land is already being farmed in the U.S. In other areas, there can only be marginal increases in the area of land farmed. Water limits the use of additional land for farming.
  - **Use drip irrigation & hydroponics** - By using water more efficiently, we may be able to extend the limited water supplied into areas that don't have sufficient water. Drip irrigation is a method of efficiently supplying just the water that is needed for a crop. Hydroponics is growing crops such as lettuce and tomatoes in a totally artificial nutrient medium. This can be done effectively, but it is costly.
  - **Eating lower on the food chain** - Some people believe that it would be better to use rangeland for growing grains for human consumption rather than to feed the grain to cattle and other domestic animals and then eat the animals. The benefit would be that less food energy would be lost to entropy in the food chain, and thus more food would be available for people. There are complicating factors with this simple idea, because not all land that is suitable for grazing can be used for grain production. Some land is only good for rangeland, like on steep slopes or in areas with low rainfall. In addition, animals will always be required by people for dairy food and meat, because humans are simply not adapted to be herbivorous.
  - **Modification of food distribution** - Using massive relief efforts, there is a way of making the food surpluses available where food is scarce. There are many problems to be overcome in this approach, not the least of which is the tremendous cost of transportation of food to where it is needed. Governments need to cooperate and continue to share food among the rich and poor nations, but this is not a long-term solution to world hunger. The solution is to increase local food production rather than to force people to rely on foreign assistance.

What foods should I consume if I want to eat a vegetarian diet but still obtain the essential amino acids?
- You may wish to consume a vegetarian diet for many reasons, including moral reasons, religious reasons, or you simply wish to "eat low on the food chain."
- Here are some things to bear in mind when and if you become vegetarian.:

- Normally humans are omnivores, not herbivores (see Schneider et al. 1977). The fossil record indicates that our ancestors ate seeds and insects, not leaves. We have largely dropped insects from our diets (except in certain cultures), but we have replaced them with meat. We have evolved from a society that was composed of hunters of meat and gatherers of fruits and seeds. Our dentition and digestive physiology reflect this evolutionary pattern, and these are different from the ruminants, which can obtain energy from cellulose. Humans can get some energy from cellulose, but most of our energy intake is from less complex carbohydrates such as sugars and starches.
- As you can see from the discussion of malnourishment above, one special dietary nutrient to be concerned about is proteins, which are made of amino acids, and are required for a healthy diet. You must consume all the **essential amino acids** (histidine, isoleucine, leucine, lysine, methionine, phenylalanine, threonine, typtophan, and valine) in order to make the proteins your body needs to function. These essential amino acids cannot be synthesized rapidly enough by your body, and thus they must be present in your diet. Vegetable proteins are often lacking in essential amino acids; animal diets are not.
- Vegetarian diets can be nutritionally adequate if the foods consumed are selected wisely, because one vegetable lacking in a specific amino acid can be complemented with another vegetable protein that has the missing nutrient.
  For example, grains and legumes should be consumed together to provide all essential amino acids. One problem for those that choose to consume no animal products (no eggs, cheese, dairy products, etc.) is the absence of vitamin B_{12} in the diet, which is only found in animal food (insects are animals, so they are rich in B_{12}). "Vegans", people who eat no animal products, must take vitamin B_{12} supplements to maintain good nutritional health. Also, they should take calcium supplements, because most calcium comes from dairy products. Vegetarians should eat foods in combination as shown in the following table. Always consume something from Columns I and II:

Column I	**Column II**
Grains: Barley, corns, oats, wheat, rice	**Legumes**: Beans, peas
Nuts and seeds: almonds, Brazil nuts, pecans, sunflower seeds, walnuts	**Dairy Products**: Cheese, eggs, milk, yogurt, ice cream

What is the Green Revolution?

- The name Green Revolution is given to the suite of crop improvements that were developed after World War II, including the use of genetic strains that grew well under conditions of unlimited fertilizer and pesticide application.
- These high-yielding varieties have produced some tremendous crops in the past, but there may be a limit to the amount of increased productivity that can be obtained using these methods.
- Also, because of the exclusive use of these genetic superstrains, overall crop genetic diversity has declined.

What percentage of the Earth's surface can be farmed?

- Only 11 % of the Earth's land surface is considered good for farming.

- In the U.S., it is more like 25 %, but 80 % of that is already being farmed. In the U.S., there are 190 million ha of farmland for crops, and another 300 million ha for pasture and grazing.

How much farm land is lost each year in the U.S.?

- There is a net loss of 500,000 ha/year. Actually, 1 million ha are lost to development for human housing each year, but there are an additional 500,000 ha that are added through wetland drainage and irrigation of dry lands.

What is aquaculture?

- Aquaculture is the production of food from aquatic habitats using controlled breeding and rearing environments. There are many species of fishes and invertebrates that can be grown in culture environments, as the Chinese have been doing with carp in rice ponds since 475 B.C.
- In this ancient Chinese practice of carp and silkworm culture, carp are grown in ponds with mulberry trees planted around the edge. Silkworms consume the mulberry leaves and their waste products stimulate pond phytoplankon, which are fed upon by the carp. The sediment from the pond is used for fertilizer, so all nutrients are recycled in this simple system.
- There is great potential for the world to obtain some of its animal protein by means of aquaculture in the future. However, except in a few situations with a few species like catfish and shrimp, this practice is too expensive or it takes too long to obtain a marketable crop. In addition, unless every waste product is recycled as in the carp aquaculture, there will be pollution problems with aquaculture.

What is mariculture?

- This is aquaculture in a marine environment.
- There are many examples of successful, small-scale mariculture projects, but only a very few are economically viable.
- Pen-raised salmon are perhaps the best example of a closed mariculture system in the sea, but these fish are a high-dollar crop that will do little to solve the world's food crisis.
- There is potential for species like tilapia, which unlike salmon is a herbivore that can be grown in salt-water pens or ponds.
- Shrimp farming is extremely profitable, and 22 % of the world's shrimp are farm-raised.
- It is unlikely that mariculture will be the answer to world food problems, because it is far cheaper and easier to capture naturally spawned wild fish and invertebrates.

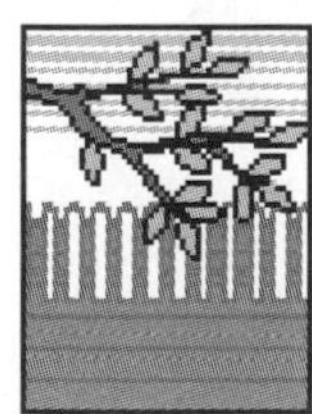

Ecology In Your Backyard

- Where does your food come from?
- What are the most common grains in your diet?
- What percentage of your diet is animal and plant food?
- Are you a vegetarian or an omnivore in your food habits?

- How can you eat lower on the food chain?

Links In The Library

- Register, U.D., L. M. Sonnenberg. 1973. The vegetarian diet. J. Am. Diet. Assoc. 62: 253.
- *Recommended Dietary Allowances*. Washington, DC, National Academy of Sciences Press.
- Schneider, H. A., C. E. Anderson, and D. B. Coursin (eds.) 1977. *Nutritional Support of Medical Practice*. Medical Department, Harper & Row, Publishers, Hagerstown, MD.

Ecotest

1. Which of the following must a vegetarian human consume to ensure that all the essential amino acids are available in their diets?
a. beans and milk
b. legumes and cheeses
c. grains and legumes
d. nuts and grains

2. The most efficient way to water crops is to use ________ irrigation.
a. ditch
b. drip
c. canal
d. spray

3. Which of the following foods is of greatest importance (by biomass) in the diet of the human population?
a. hamburger meat
b. chicken
c. fish
d. wheat
e. none of these is correct

4. Which is greater on an annual basis, the amount of food produced per person, or the amount of food needed per person to meet basic needs?
a. food produced per person
b. amount of food needed per person
c. The amounts are the same.

5. What is the main cause of world hunger?
a. There is not sufficient food produced each year in the world.
b. Sufficient food is produced, but it is poorly distributed in the world.
c. There is sufficient food, but hungry people are too weak to feed themselves.
d. There is no world food problem.

6. ______________________ is caused by a diet of too few calories.
a. Marasmus
b. Kwashiorkor
c. Undernourishment
d. Malnourishment

7. Food production, measured in number of days of grain supply on-hand, has been
a. holding steady at 104-day supply since 1973.
b. steadily declining since 1960's high of a 62-day supply.
c. steadily increasing since the 1960 low of a 56-day supply.
d. fluctuating, but has declined since 1970's high of 104-day supply to the 1995 level of a 62-day supply.

8. Agro-ecosystems differ from normal ecosystems in which of the following ways?
a. They are late-successional stages.
b. The are polycultures of many crop species.
c. They have high habitat complexity.
d. They have simple food webs and food chains.
e. They have infrequent soil disturbance.

9. ______________________ agriculture is done on small patches of cleared forest land with many species of crops grown together, without any rows, pesticides, or fertilizers.
a. Swidden
b. Milpa
c. Fang
d. Bush fallow
e. All of these are correct.

10. One sustainable method of aquaculture that has produced food for humans for over 2000 years and that is becoming more common today is ____________________________.
a. farming of salmon in pens
b. shrimp ponds
c. tilapia culture
d. carp culture

11. Of all the species of plants present on Earth, what percent are grown as agricultural crops?
a. 50 %
b. 25%
c. 10%
d. less than 1%

12. Examples of a monoculture include all of the following except:
a. pine plantations.
b. golf course putting greens.
c. stocked catfish farm ponds.
d. rainforest.
e. cotton fields

13. Planting crops in a sequence of places over many years to allow soil nutrients to recover or be replaced is called:
a. crop circles.
b. disturbance.
c. crop rotation.
d. succession.

14. When soils lack a specific nutrient required for plant growth, the nutrient is said to be a/an:
a. macronutrient.
b. trace element.
c. amino acid.
d. limiting factor.

15. Agricultural production can be increased by all of the following except:
a. increasing the area of land that is farmed.
b. using water more efficiently.
c. eating lower on the food chain.
d. modifying how food is distributed.
e. raising the price of food products

Chapter 12 - EFFECTS OF AGRICULTURE ON THE ENVIRONMENT

The Big Picture

The landscapes of the arable portions of the Earth have been greatly altered by agriculture. Centuries ago, when the global human population was considerably smaller than today, the conversion of the natural landscape (e.g., forests and grasslands) to agricultural landscapes was localized. The eastern Mediterranean, portions of northern Europe, and population centers of eastern Asia were drastically altered by grazing or intensive crop production. Today, human populations are exponentially greater and extend to most of the inhabitable portions of the planet. The effects of agriculture are not only local but global, as well. Agricultural production is necessary to maintain modern human cultures but there are serious environmental trade-offs. Agricultural production entails the production of plants and animals for food and fiber. Agriculture seeks to promote the growth of species of plants and animals that provide these services. The most efficient way to maintain or increase agricultural production is to produce crops and livestock in monocultures or low-diversity systems. Thus, we are replacing ecosystems with relatively high biodiversity with agricultural ecosystems of relatively low biodiversity. Similarly, species that are not valued as agricultural commodities or interfere with the efficient production of agricultural species may be considered competing species or even pest species. In order to maximize agricultural production and minimize economic expenditure, modern agriculture in the United States and other more developed nations often relies heavily upon the use of pesticides, fertilizer supplements, and mechanized production. Using these techniques, the agricultural sector of our society is able to produce enough food for millions of people in non-agricultural sectors (i.e., manufacturing, industrial, governmental, and services). Agricultural production in the less developed nations still relies heavily upon human energy input, but large tracts of land are often necessary to satisfy demands for food and fiber production. Most farmers have long recognized the importance of protecting soil, water, and biological resources and many are good stewards of the land. Still, as the demands for agricultural products increase the stresses on natural resources will be exacerbated. Only through a combination of governmental regulation, incentive programs, and personal initiative by farmers can natural resources be sustained.

Frequently Asked Questions

What are examples of on-site and off-site effects of agriculture?

- **On-site environmental effects** are those which take place on or in the immediate vicinity of the agricultural activities. These effects include: the conversion of land to agricultural production, thus a loss of biodiversity and habitat; erosion and soil loss; soil contamination; and possible degradation of water quality and reduction of water quantity. These on-site effects would be considered local effects.

- **Off-site environmental effects** are those which occur downwind or downstream from the agricultural operation. Off-site effects may be expanded to include those effects that result from the combined effects of farming practices in a region, such as aquifer depletion or the loss of habitats. Globally, agricultural production has the potential to alter chemical cycles and climatic conditions.

How are soils affected by agricultural production?

- There are over 17,000 soil types recognized worldwide. They vary widely in structure, erodibility, fertility, and ability to produce crops. A generalized soil profile for a humid, temperate climate is depicted in Figure 12.3.
- When the natural vegetation is cleared for agriculture, soils become exposed to erosion and loss of soil fertility.
- The removal of the above-ground natural vegetative cover and subsequent plowing and preparation for row crop planting also destroys plant roots which would otherwise help stabilize the soil. Soils disturbed by plowing and cultivation are prone to erosion by water runoff and wind. The erodibility of a particular soil depends upon soil type, topography (steepness), organic matter, and intensity of the erosive processes.
- In the United States, soil erosion is estimated to be as severe or worse today as during the Dust Bowl years of the 1930s (Figure 12.4). One-third of the nation's topsoil has been lost to erosion.
- Much of the eroded sediment eventually are deposited in streams, rivers, lakes, and the ocean. In the United States, approximately 4 billion tons of eroded sediment is deposited in waterways each year and 75% of this originates from agricultural lands.
- When the sediments enter waterways, habitat quality for aquatic plants and animals may decline, as well as water quality for human use.

How can we reduce soil erosion?

- If farmland is allowed to recover by allowing it to remain fallow and letting natural succession take place or by actively restoring the original vegetation type, soil-forming processes (pedagenesis) will proceed. However, soil forming processes can be very slow.
- Removing highly erodible land from production is one strategy to reduce soil erosion. State and federal incentive programs are in place to promote this strategy.
- Obviously, much farmland must remain in cultivation to satisfy the demand for agricultural products.
- A variety of plowing and cultivation techniques can be used to reduce soil erosion. These include contour plowing (following topographic contours) and no-till agriculture (minimal disruption of soil surface). Other strategies include timing plowing and cultivation to time when the potential for erosion is reduced, crop rotation, planting multicultures, strip-cropping, terracing, and "grassed" waterways (these are drainage ditches with grass cover) (Figure 12.7).
- The success of the techniques depends on local soil conditions and the types of crops being planted.

How does agriculture affect soil fertility and when do plant nutrients pose an environmental problem?

- Nutrients cycle between soil, water, air, and the biota (see Chapter 4). Agricultural production interferes with the rate of nutrient cycling and ultimately redistributes nutrients, depleting soils of some nutrients and concentrating nutrients in eroded sediments and waterways.
- Many crops have high nutrient demands, thus deplete the soil of nutrients at a faster rate than native plants.
- Soil cultivation disturbs the soil and increases the rate of nutrient loss via erosion.
- Harvesting crops removes nutrients from the system, thus they are not recycled.
- Nitrogen and phosphorus are nutrients that crops require in large amounts but that are readily depleted from the soil. If soils are of low fertility, farmers have to artificially supplement soils with inorganic and/or organic fertilizers. (Commercial fertilizers used in the United States typically contain nitrogen (N), phosphorus (P), and potassium (K), as well as inert materials and possibly trace nutrients. The numerals appearing on a bag of fertilizer correspond to the percentages of N-P-K.)
- Intensive livestock production in feedlots and concentrated animal production facilities tends to concentrate nutrients in animal wastes.
- An excess of nutrients applied to cropland or concentrated in animal wastes can result in off-site environmental problems. Through erosion and water movement, nutrients may be transported by surface waters and groundwater. Eutrophication and water quality degradation often result from nutrient mismanagement (see Chapter 21).

What are some of the environmental effects of grazing?

- In arid (desert) and semi-arid (grassland) regions where rainfall is insufficient to support row crop production, land is often used for grazing (Figure 12.15). In the United States, 40% of the land area is classified as rangeland.
- The grasslands of the Great Plains were well adapted to grazing by migratory bison and other ruminants. Today, bison have been replaced with cattle, sheep, and other non-migratory animals that forage differently than bison, thus affect plant community structure and composition.
- Overgrazing by livestock alters the plant community by removing species that are intolerant of intensive grazing and promoting the proliferation of plant species that are unpalatable. Additionally, livestock are often directly or indirectly responsible for the introduction of exotic plant species that may become pest species.
- Landscapes that were once forested have been converted to rangeland. In eastern North America, Europe, Southern Africa, and, increasingly South America and Central America, forests have been cleared and converted to pasture (Figure 12.14). This conversion results in a loss of habitat and biodiversity, and a change in ecosystem structure and function.
- Game ranching is an alternative to conventional grazing practices. Native or other "wild" herbivore species are allowed to free-range and are harvested at a sustainable rate. This non-conventional practice is controversial and would have many cultural and economic obstacles to overcome.

What is desertification and how can it be prevented?

- Deserts account for approximately one-third of the total land area on earth. The five major deserts lie between 15 and 30 latitude, north and south.
- As a result of historic overgrazing and inappropriate land use practices, some desert areas are expanding into surrounding landscapes and other semi-arid regions have been and continue to be converted to desert. This process is called desertification.
- Additional desertification may be prevented by monitoring vegetation, water table, surface and soil moisture, soil organic content, and soil salinization.
- The restoration of lands that have experienced desertification will entail reducing grazing pressure and re-establishing vegetation and habitat.

How are pests and weeds controlled in modern agricultural practices?

- "Pest" is a general term referring to any undesirable species of plant, animal, or microbe. Agricultural pests include any organisms that compete with or otherwise damage crops and livestock.
- The goal of agriculture is the production of economically valuable crops and livestock in an efficient manner. Pests reduce efficiency, thus there is considerable effort and monetary expenditure toward combating pests.
- In the U.S., economic losses of $16 billion/year can be attributed to plant pests (weeds). Weed control costs about $3.6 billion/year.
- Following World War II, sophisticated pesticides with complex chemical structures were developed. Pesticide usage in the United States and worldwide increased dramatically. In the United States today more than 300kg of pesticides are used per year (661 million lbs/yr).
- Pesticides include bactericides, insecticides, herbicides, fungicides, and rodenticides.
- **Narrow spectrum pesticides** target specific pest species that are considered less harmful to the environment. **Broad spectrum pesticides** affect a wider range of organisms including pests species as well as non-pest and beneficial species.
- The rapid reproduction rates of many pest species may allow these species to evolve resistance to pesticides; this is called **pesticide resistance**. Consequently, new and more toxic pesticides must continually be developed.

What are some of the environmental effects of pesticide usage?

- Unfortunately, many pesticides (even narrow spectrum pesticides) have environmental effects beyond their targeted use.
- Non-target species may be eliminated by pesticide use. These species may include such beneficial species as pest predators and plant pollinators.
- Some pesticides are not easily biodegraded in the environment. Long after their target use, these pesticides may persist in the tissues of plants and animals or in sediments, then re-enter the food web.
- One of the more infamous examples of the environmental side effects of pesticide use is DDT, a chlorinated hydrocarbon that was widely used in the United States as an insecticide until its use was discontinued in the 1970s.
- DDT was a fat-soluble compound that did not readily decompose in the environment. If consumed by animals, DDT concentrated and persisted in the oils and fats of the

consumer organism. Then DDT could **biomagnify** and become transferred through food chains. Organisms at the top consumer levels of food chains and long-lived organisms tended to have the highest concentrations of DDT in their tissues.

- Aquatic food webs were especially vulnerable to the harmful effects of DDT. Long aquatic food chains resulted in high concentrations of DDT in the tissues of top consumers, such as pelicans, ospreys, and bald eagles. These birds experienced damage to their reproductive systems and consequently they produced thin, easily damaged egg shells. Populations of these species declined significantly.
- DDT is no longer used in the United States, however, chemical manufacturers are legally allowed to produce DDT and sell it overseas. DDT is often used in the tropics to control mosquitoes and other disease vectors.

What is Integrated Pest Management (IPM)?

- IPM is a common sense strategy to manage pests, minimize environmental damage, and maximize farm profits.
- IPM utilizes a suite of possible solutions rather than relying solely on a single method of pest control, such as relying heavily on pesticides (Figures 12.9 and 12.10).
- IPM strategies include using pesticides judiciously and only after determining 1) the specific nature of the pest problem, 2) the proper application rates, and 3) the proper timing of application.

Ecology In Your Backyard

- What are the principal agricultural products of your region?
- Are there any specific environmental problems associated with agricultural production in your region?
- What alternative agricultural practices are being used or may be used to ameliorate any harmful effects?
- Have any environmental or human health problems associated with agriculture received media attention?
    - **For example**: Hog production in eastern North Carolina is an example of how a very intensive agricultural practice can have major off-site effects on the local and regional environment, and on human health and well-being. A Pulitzer Prize winning series of newspaper articles in the Raleigh, NC News and Observer describes the effects of the hog industry on air and water quality. Since the publication of the newspaper series, the hog farming issue has become much more acute. In the summer of 1995 and fall 1996, above average rainfalls resulted in saturated soil conditions and failure of the waste lagoon and spray field disposal system. In June 1995, millions of gallons of hog waste were inadvertently discharged into coastal rivers. In one instance, 25 million gallons of hog waste

(more than twice the volume of contamination from the Exxon Valdez) flowed into a coastal river.)

- *Soil Surveys* (Natural Resources Conservation Service) have been published for most counties in the United States. These may be obtained from county extension agent offices or NRCS offices for a nominal fee or cost-free, or see your local or university library. Soil Surveys contain a wealth of information pertaining to local soil types and suggested land uses based on soil type.

Links In The Library

- Briggs, S.A. 1992. *Basic Guide to Pesticides: Their Characteristics and Hazards.* Taylor and Francis. Washington, D.C. 283 pp. This book, produced by the Rachel Carson Council, provides an overview of pesticides and their toxicity. Common, trade, and chemical names of pesticides are cross-indexed. An extensive chart, indexed by common name, lists the class, status, persistence, and adverse effects of each pesticide. Appendices include discussions of pesticide usage, carcinogenity, environmental and economic effects, federal legislation, and alternatives to pesticides (IPM).
- Dunlap, T.R. 1981. DDT- *Scientists, Citizens, and Public Policy.* Princeton University Press, Princeton, NJ.
- National Research Council. 1989. *Alternative Agriculture.* National Academy Press, Washington.
- USDA-NRCS *Soil Surveys* may be obtained for your county at the local NRCS office or your library.
- *Farm Journal* - A popular farming journal.
- *Progressive Farmer.* Farming strategies, conservation, pesticide use, farm economics, and governmental regulations pertaining to farming are discussed.

Ecotest

1. The soil horizon composed of a mix of organic and inorganic materials lying directly below the organic horizon is the _____ horizon. It is commonly referred to as topsoil.
a. A
b. B
c. C
d. D

2. Approximately _____ of the topsoil in the United States has been lost due to erosion.
a. three-fourths
b. one-half
c. one-third

d. one-quarter

3. Plowing and cultivation techniques that help reduce soil erosion include all of the following except ______.
a. contour plowing
b. pedegenesis
c. strip-cropping
d. no-till agriculture

4. Standard commercial fertilizers used by farmers and gardeners contain three of the following nutrients. Which of the following is not a plant nutrient supplied by fertilizer?
a. nitrogen (N)
b. potassium (K)
c. silica (SiO_2)
d. phosphorus (P)

5. What percentage of the United States may be classified as rangeland?
a. 20%
b. 30%
c. 40%
d. 50%

6. ________ is the conversion of non-desert land to desert, usually as a result of inappropriate land uses such as overgrazing.
a. Soil salinization
b. Sclerophication
c. Desertification
d. Aridification

7. Which of the following types of pesticide would be expected to cause the fewest environmental problems?
a. a pesticide that biomagnifies
b. a fat-soluble pesticide
c. a broad spectrum pesticide
d. a narrow spectrum pesticide

8. A pest control strategy that utilizes a suite of pest management techniques such that environmental damage is minimized is called _______.
a. Integrated Pest Management (IPM)
b. Alternative Pest Control (APC)
c. Broad Spectrum Management (BSM)
d. Strategic Defense Initiative (SDI)

9. Which of the following statements about DDT is false?
a. DDT is a chlorinated hydrocarbon used as an insecticide.
b. The manufacture and use of DDT in the United States has been banned since the 1970s
c. DDT presents a great risk to top predators in aquatic food chains.
d. DDT is a broad spectrum pesticide.

10. Which of the following statements about agriculture in the United States is false?
a. Pesticide resistance is an on-going dilemma.
b. The historic and current conversion of landscapes into extensive monocultures reduces biodiversity.
c. Sophisticated pest control techniques have virtually eliminated crop losses due to plant pests.
d. Soil erosion is as severe today as it was during the 1930s.

11. Which of the following is an example of an off-site effect that a farm could have?
a. Soil becomes contaminated with salts and heavy metals from irrigation.
b. Native species of grasses die because of intolerance to high influx of water.
c. Town reservoir decreases in depth because so much water is being used by the farm.
d. Irrigation ditches between crop rows fill with sediment.

12. There are over _________ different soil types recognized worldwide.
a. 100
b. 1,800
c. 17,000
d. 1 million

13. The major deserts are found:
a. at the North and South Pole.
b. throughout all parts of the world.
c. in the temperate parts of the world.
d. within 15 to 30 degrees of the equator.

14. Deserts account for what fraction of the Earth's surface?
a. 1/2
b. 1/3
c. 1/4
d. 1/5

15. The most efficient way to maintain or increase agricultural production is to produce crops and livestock in:
a. multispecies assemblages.
b. monocultures.
c. the species native to the system.
d. competition with other organisms.

Chapter 13 – FORESTS, PARKS, AND LANDSCAPES

The Big Picture

The Earth's surface is partly land and mostly water. This chapter discusses the Earth's terrestrial ecosystems (landscapes) and the forest resources they contain. The Earth's **landscapes** may be classified into many land use categories, including forested areas, pasture, agricultural land, deserts, wetlands, and urbanized areas. Within the world's forested lands, there are different forest types, including rainforests, dry forests, coniferous forests, deciduous forests, closed canopy forests, and open canopy forests. Forests are valued greatly by humans, as they are used for lumber, boat-building, and firewood. Forests also regulate climate, store water, prevent erosion, store carbon, provide wildlife habitat, and provide recreational opportunities, in addition to producing trees. These functions are important to humans, but are often not valued by humans to the same extent. Forests can be managed to maximize tree production, such as on a pine plantation, but if a forest is managed for only one function, other functions may be compromised. Losses of forest resources are occurring very rapidly worldwide. This is a cause for concern among many environmental groups, because of the loss of these functions and values (see the Case Study on Conserving Wetland Forests). Changes in landscapes (deforestation, increased run-off from agriculture and urban areas) are causing impacts in the coastal seas.

Frequently Asked Questions

What is the Landscape Perspective?

- In most cases, conservation of natural resources requires that management actions take into account processes in nearby ecosystems, not only in the one being managed.
- This larger view is the landscape perspective.
- Management at this landscape perspective requires knowledge at several spatial scales – local to regional – with many kinds of ecosystems.

How should forests be used?

- Some think that forests should be a source of wood and forest products.
- Others think that forests serve a larger function as wildlife habitat, and provide natural ecological services.
- Forest management is a controversial topic, because different people have different ideas about how forests should be used.

How much forest land is there in the world?

- Today, there are 3.45 billion hectares of forest worldwide (1 hectare (ha) = 2.47 acres).
- This is a decline from 1980, when there were 4 billion ha worldwide.
- Forests cover 26.6 % of the Earth's Surface.
- See Table 13.2

How much forested land is left in the U.S.?

- There are 212 million hectares of commercial-grade forests in the U.S., with 75 % of them in the East (especially the pine plantations of the Southeast), and 25% in California, Oregon, Washington, Montana, Idaho, Colorado, and Alaska.
- 70% of commercial-grade forested land is privately owned, 15% is on U.S. Forest Service land, and 15% is on other federal lands.

How do trees function in ecosystems?

- There is a great diversity of tree species.
- The greatest tree diversity is in the tropical rainforests.
- Common functions that trees perform in ecosystems: photosynthesis, transporting water and nutrients, and providing structure in ecosystems
- The diversity of tree species in forests is due in part to niche specialization. Some trees are shade-tolerant, others are fire-adapted, others are drought-resistant.
- Forests reflect sunlight, increase evaporation rates, absorb greenhouse gasses, reduce windspeed.

What is meant by sustainable forestry?

- The phrase "sustainable forestry" can mean different things and is used differently by different people. There are two definitions of sustainable forestry:
    - **Sustainable timber harvests** means that wood production may be sustained for an indefinite period of time under a maximum sustainable yield model on a tree plantation; however, other forest ecosystem functions may not be sustained under such a plan.
    - **Sustainable forest harvest** means that forest ecosystem functions are sustained for an indefinite period of time; these include carbon storage, soil erosion protection, climate regulation, water storage, wildlife habitat, and recreational opportunities for people.
- Sustainability of forest functions has not been experimentally demonstrated, except when harvest rates are kept very low (**coppicing**, which is the selective removal of limbs from trees, and selective cutting, but not clear-cutting). It is quite possible that when a person says "sustainable forestry" they could be referring to either of these meanings.

What is certified forestry?

- When forest managers demonstrate that they are harvesting forests in a way that is consistent with sustainable forestry (they replant trees that have been cut and they maintain buffer strips along streams, etc.), they can be certified as sustainable.

- However, there is disagreement among foresters as what is sustainable forestry.

Are currently used forest harvest practices sustainable?
- During the past 100 years, forest harvest rates have been higher than rates of regrowth and recovery.
- During the whole of human history, there has been a net loss of forested area worldwide and in the U.S. Thus, the practice of forestry as it is practiced today is non-sustainable over the long-run.

What is silviculture?
- Growing trees for harvest.

What is an "old growth" forest?
- This term means that the trees in the forest are very old and have never been cut in the last 250 years. Often, the trees are 1000 years old or more.
- The term "ancient forest" or "virgin forest" is often used as an alternative.
- These three names are not scientific terms, but are used in many well-publicized disputes about logging in forests in the Pacific Northwest of the U.S.
- The term **second growth** refers to a forest that has been cut and is regrowing.

How are commercial forests currently managed?
- There are two major types of management of forests: plantations and natural regrowth.
- **Plantations** (tree farms) are managed for either maximum sustainable yield (MSY) of wood for lumber or a short rotation time for wood pulp for paper production with seedlings planted after a clear-cut, and fertilizer, pesticides, and herbicides added as needed, sometimes by helicopter (Figure 13.12, early successional stage). These are typically even-aged stands, with all the trees the same age.
- **Natural regrowth** management is less intensive than a plantation and allows trees to regrow from seeds deposited by neighboring adult trees that were left intact after selective cutting and natural succession are allowed to occur. These forests will have uneven-aged stands of trees, and some trees will be old and useful for making hardwood furniture (Figure 13.13, late successional stage).

Where is most deforestation occurring today?
- In the developing world, which lost 200 million ha in 1980-1995.
- Consequences are increased soil erosion, stream sedimentation.

What are the causes of deforestation?
- There are two kinds of deforestation: direct and indirect.
- **Direct deforestation** is caused by cutting trees for industrial forest products (paper and lumber), fuelwood (cooking and heating), and to clear land for agriculture (cattle ranching, palm oil plantations, tobacco farms). Direct deforestation occurs worldwide, but is occurring most rapidly in tropical South America, Asia, and Africa.

- **Indirect deforestation** is due to air pollution by acid rain and ozone, which in combination with insects and parasites, kills the trees. Indirect deforestation is especially a problem in Europe.

What is clear-cutting and what does it do to a forest ecosystem?
- Clear-cutting is a practice in which all the trees in a stand, no matter what species, are cut down and harvested over large areas of forest, leaving behind stumps or unvegetated soil.
- No trees are left to re-seed or re-vegetate the land, which must be replanted by humans with seedlings.
- Clearcutting increases soil erosion and runoff, causes nutrient cycling to be disrupted, increases nitrate levels and sedimentation in streams (Figure 13.15).
- Experimental tests of clear-cutting have been done in the U.S. Forest Service's Hubbard Brook experimental forest in New Hampshire and the H. J. Andrews experimental forest in Oregon.
    - When an entire watershed was clear-cut, and compared against a nearby watershed that was left forested, there were dramatic increases in soil erosion rates and stream sedimentation rates. At the Oregon forest, landslides occurred after clear-cutting.
    - In addition, water storage declined and nutrients were lost from the ecosystem. Nitrate rates in the watershed's central stream exceeded public health standards in the New Hampshire clear-cut forest.
    - The increased stream sedimentation, landslides, and water nutrient cause spawning of anadromous fishes like salmon to decrease. This is why fisherman are opposed to clear-cuts near salmon streams in the Pacific Northwest.

What are the alternatives to clear-cutting and what are their benefits and drawbacks?
- The main advantage of clear-cutting to timber companies is the low cost to harvest the trees. There are several alternatives to clear-cutting, but all of these involve cutting fewer trees per unit area. Thus the major drawback of each alternative method is the increased cost of harvesting trees. The alternatives range from almost no cutting (selective cutting) to some that are similar to clear-cutting (seed tree cutting). Here they are in order of most desirable to least desirable alternatives to clear-cutting:
    - **selective cutting** - Small groups of individually marked mature trees are cut and removed. The rest of the forest remains intact, and the forest can regenerate naturally. Benefits include a natural seed source, little or no soil erosion, and a minimal impact on wildlife habitat.
    - **shelterwood cutting** - In this method, all the mature trees are harvested over a period of time. In the first year, undesirable trees are removed. Afterward, new desirable tree seedlings become established. Then after a decade, many mature trees are removed. After another decade, the last mature trees are cut, but by this time there are many young trees regenerating. Benefits of this

method include natural reseeding and low soil erosion, because some trees are always present.
- **strip cutting** - In this method, long, wide swaths of forest are removed. This allows trees in the non-cut swath to reseed and protect the new seedlings from excessive sun and wind. It also provides wildlife habitat and an aesthetic barrier from the strip that has been cut. However, soil erosion can be high in cut strip.
- **seed-tree cutting** - In this method, all the trees except for a few scattered mature trees are cut. The mature trees left behind produce seeds for regeneration. Benefit is seed production alone; the soil is largely exposed as in a clear-cut, so that erosion may be high.

How can parks and preserves protect landscapes?

- Parks and preserves protect forest and other landscape ecosystem functions by preventing extractive use of the land. This approach has gained a great deal of favor in recent years among conservation biologists, using Island Biogeography Theory.
- **Parks** provide people with recreational areas, protect biodiversity, provide water storage, provide wildlife habitat, and protect places of aesthetic value (Yosemite National Park is such a place).
- **Nature preserves** are somewhat different than parks, in that they are solely established to protect nature or wilderness for its own sake, not for humans. In some parks and preserves, humans are not allowed (for example, Sengwa National Park in Zimbabwe) so that natural ecosystems can exist without human influence.

How much land should be in parks and preserves?

- Parks and preserves are essentially "islands" surrounded by an altered landscape in much the same way that oceanic islands are surrounded by water. Thus, the principles of this mathematical theory can be applied to park and preserve "islands".
- Generally the larger the island, the more species it will contain, and the closer the island is to a source of species the more species it will contain. The same is true for parks and preserves. If a park is large, this is generally thought to be better than a small area park, because it will contain more biodiversity.
- However, several small parks of the same total area can often have as many species as one large park, with an advantage of guarding against catastrophic loss in one of the parks (Figure 13.18).
- Also, it is good to have a series of parks, or to have parks reside close to large areas of wilderness that can serve as a source of species.
- Park boundaries need to be drawn with the wildlife that use the area in mind. Often, species' natural ranges extend beyond the park or preserve boundary and the species can be impacted or conflict with humans outside those areas.
- Wildlife corridors between parks and nature preserves often function to allow the animals to interbreed and migrate among parks and are thought to increase the genetic diversity of the populations within the parks.
- About 10 % of the total land area of a country has been recommended, although there is no definitive amount that has been shown to be better than others.

- Costa Rica and Kenya have placed 10% or more in parks or reserves. France has 7% in various regional parks, but only 0.7 in national parks.
- The U.S. has 10.5 % in parkland and wild areas (98 million ha, mostly in the western states). Some U.S. states have very little wilderness remaining, however.

What is an edge effect?

- This occurs when an "island" of forest is created and a disturbance zone is created along the edge of the new "island" of forest. Along the edge of the new isolated patch of forest, the area is disturbed, there is more light, more wind, and less protection. Some species disappear in this disturbed zone, and others that survive well in disturbance areas colonize the edge habitat.
- There is a net loss of species from the cut area when it is cut off from the main forest, and an increase of those species in the adjoining areas. This is called the edge effect. If the island of forest is too small, the entire area may be influenced by the edge effect.

What is a wilderness area?

- A wilderness area is a roadless area that is undisturbed by humans. If humans are present, they are visitors, not residents. In the U.S. there is a law (Wilderness Act of 1964) that specifically protects wilderness.
- The law designates wilderness as "...an area of undeveloped Federal land retaining its primeval character and influence, without permanent improvement or habitation, which is protected and managed so as to preserve its natural conditions." In addition, it must be larger than 5000 acres (2025 ha).

Ecology In Your Back Yard

- What kind of wood is in your home?
- Where does the wood in your home come from?
- How much paper do you use every day?
- Where does the paper you use come from? Is any of it recycled?
- How can you reduce paper usage and thus save trees?

Links In The Library

- Hunter, M.L. 1990. *Wildlife, Forests, and Forestry: Principles of Managing Forests for Biological Diversity*. Prentice Hall, Englewood Cliffs, NJ.

- Perlin, J. 1989. *A Forest Journey: The Role of Wood in the Development of Civilization.* Harvard University Press, Cambridge.
- Perry, D. A. 1994. *Forest Ecosystems.* Johns Hopkins University Press, Baltimore. 649 pp.

Ecotest

1. Clear-cutting a forest can result in:
a. the selective removal of certain tree species.
b. an increase in wildlife diversity and abundance.
c. severe levels of erosion and fish population declines in the streams of the surrounding watershed.
d. the increased productivity of trees planted to replace those removed.

2. One advantage that clear-cutting a forest has over selective cutting is that it:
a. sustains species diversity.
b. is less disruptive to ground cover and soil.
c. is more economical.
d. produces sustained yields.

3. The term ____________refers to a forest that has not been cut in the last 250 years or more.
a. old growth
b. second growth
c. canopy forest
d. plantation

4. What is an "edge effect?"
a. The area around a habitat boundary where some species can live, but which does not provide the same protection as in the center of a habitat.
b. In strip cutting method of forest harvest, the impact of deforestation only affects the edge of the forest ecosystem.
c. When a track is cut by a skier or a mountain-biker, the edge of the track promotes more species and higher biodiversity.

5. In the creation of nature reserves, conservation biologists must consider:
a. All of these are correct.
b. the home range size of the species to be protected.
c. island biogeography theory.
d. the specific habitat requirements of the species to be protected.

e. inbreeding depression.

6. As a forest management practice, clear-cutting:
a. will always lead to major loss of soil nutrients in run-off.
b. may be important in inducing regeneration of desirable species.
c. is not used much anymore.
d. is usually best used on hill slopes.

7. "An area of undeveloped Federal land retaining its primeval character and influence, without permanent improvement or habitation, which is protected and managed so as to preserve its natural conditions" is the legal designation for a:
a. historic site.
b. national park.
c. wilderness area.
d. seascape.

8. Which of the following is not an example of indirect deforestation?
a. Removing trees to create agriculture fields.
b. Insect parasitism resulting in the death of an old-growth stand of trees in a pristine area.
c. Nutrient leakage from an industrial hog farm that results in toxic levels of nutrients in a swamp forest.
d. Increased acid rain as a result of a pulp mill moving into the area causing trees to be more susceptible to a fungal parasitic blight.

9. The amount of land that has been set aside in parks and reserves to sustain and protect the landscapes in the U.S. is approximately:
a. 5 %
b. 10 %
c. 15 %

10. What is a second growth forest?
a. One that has never been cut.
b. One that has been cut once and is regrowing
c. One that is replanted on a thirty year rotation.

11. What percentage of land is forested today?
a. 10%
b. 50%
c. 26%

12. _______________ is a management practice that takes several spatial scales into consideration.
a. Coppicing
b. Forest certification
c. The landscape perspective

13. The profession of growing forest trees is called _______________.
a. silviculture
b. farming
c. horticulture
d. mariculture

14. Where is most of the deforestation occurring in the world today?
a. The USA, Pacific northwest
b. Developing countries in the tropics
c. Siberia and Russia
d. Antarctica
e. New Jersey

15. Which of the following is a function of trees in a forest ecosystem?
a. Increases evaporation
b. Photosynthesis
c. Reduces greenhouse gasses
d. Reduces windspeed
e. All of the above

Chapter 14 – WILDLIFE, FISHERIES, AND ENDANGERED SPECIES

The Big Picture

The wild living resources on Earth include wildlife, fish populations, plants, and other biota. These species may be harvested by humans for use as food, in making drugs, and as articles of clothing and fashion. Other species simply provide ecological services, such as scavenging, decomposing, acting as a predator, or as a producer. Some of these species are very abundant and other species are uncommon. Some may be on the brink of extinction; these are endangered species. Abundant species that are harvested by humans can be managed using mathematical models initially developed by foresters to determine the rate of maximum harvest. The harvest that will produce the largest yield over the time period specified is determined with a Maximum Sustainable Yield (MSY) model. These MSY models are now used by natural resource managers for fish and wildlife populations. Management of populations that are on the brink of extinction requires a different approach than management of the abundant species. Endangered species must be protected using special recovery programs that preserve the species' habitat and remove human hunting, fishing or other influences. In some cases, when the species population is so reduced that there are just a few remaining and its natural habitat is altered, intensive captive breeding programs may be undertaken so that the few remaining individuals can be carefully monitored and allowed to reproduce without interference. Any new offspring can be reared away from predators or dangers from human activities in the natural habitats that may kill them. The Case Study on Two Geese shows that the recovery programs for endangered species vary in success: the Aleutian goose has recovered from extinction, whereas the Hawaiian goose has not. In this chapter, the authors describe the causes of extinction, compare modern extinction rates to past rates, examine the traits shared by many endangered species, and demonstrate that the risk factors for extinction are increased by human technology and the human population's size. In addition, the concepts of carrying capacity, MSY models, minimum viable population sizes, minimum viable habitat, and other approaches used in wildlife management will be discussed.

Frequently Asked Questions

What is conservation biology?

- Conservation biology is the applied science of deciding what species to protect, determining what their habitat requirements are, developing strategies to protect them, and using various habitat conservation plans and technologies to prevent species from going extinct.

- Goals of conservation biology are to do at least one of the following for an endangered species (in decreasing order of preference):
  - To protect breeding populations of the species in a wild habitat.
  - To protect breeding populations of the species in a managed habitat.
  - To preserve a population of the species in a zoo so that the genetic characteristics are maintained.
  - To preserve the genetic material (frozen cells) only so that the species may later be cloned.

Why protect endangered species?

- Species cannot be replaced if they go extinct. Thus, we may want to have the species later on, even if we don't know how to value it right now. (As Aldo Leopold once wrote: "....to keep every cog and wheel is the first precaution of intelligent tinkering.")
- **Ecological Justification** - We should protect species because of their roles in ecosystems as producers, predators, decomposers, scavengers, and keystone species. We should assume that every species has an ecological role and thus an ecological value, even though we may not understand the role yet.
- **Utilitarian Justification** - We should protect species because they produce a useful product for humans (desirable genetic characteristics, medicine, food, clothing, shelter, tools, fuel). Tourism is such a useful product (people pay a lot of money to view endangered species and biological diversity; consider whale-watching and rainforest tours). Pollution control is accomplished with some species of microbes, plants, and animals; for example, bioremediation is a process of using bacteria to clean up toxic chemical spills.
- **Aesthetic and Cultural Justification** - We should protect species for their aesthetic value and their beauty. Certain majestic species, such as the American bald eagle, are protected in part because of an aesthetic and cultural value; the eagle is a national symbol of strength in the U.S. Because many people will visit beautiful regions with such species, preserving beautiful endangered species can lead to tourist dollars in the local economy (utilitarian justification).
- **Moral and Ethical Justification** - Species have a moral right to exist beyond any human needs. If we consider humans to be the stewards of the Earth and all the life on it, then we have a moral and ethical commitment to protect all species, independent of our need for them.

How many endangered species are there?

- There are 5876 threatened and endangered animal species on the International Union for the Conservation of Nature (IUCN) list.
- This total combines the species in the categories Endangered (1182 species), Vulnerable (1479 species), Rare (1144 species), and Indeterminate (2070 species).

- The IUCN reported, in 1997, that 33,798 species of vascular plants were either recently extinct or endangered; this amounts to 12.5% of all known species and 9% of the known tree species.
- The number of individuals remaining in each species varies by species and is not known precisely for any species.
- There are more endangered mollusks and insects than any other group of species of animals (See Table 14.1 in text).

What does the Endangered Species Act do?

- The Endangered Species Act of 1973 made it illegal to trade in endangered species or products made from them.
- It also made it illegal to harass, harm, or capture an endangered species or destroy its habitat.
- All federal agencies were required to prevent further loss of endangered species by using their authorities to conserve endangered species and their ecosystems.
- The act authorizes the U.S. Fish and Wildlife Service to create endangered species recovery plans, which can restrict development planned for habitats of endangered species.
- The Act declared that endangered species of plants and wildlife "...are of aesthetic, ecological, educational, historical, recreational, and scientific value to the nation and its people."

What are some common traits of endangered species?

- Many are vertebrates
- Low reproductive rates (1-2 offspring per 2 years)
- Long-lived (50 or more years)
- Large body size
- Large home range
- High amounts of food required per individual
- Top of food chain carnivores (wolves, sharks)
- Large herbivores (elephants)
- Specialist species (i.e., panda bears only eat bamboo leaves)

What is the difference between threatened and endangered status?

- Under the Endangered Species Act, a **threatened species** is one that is in danger of going extinct in part of its natural range, but is still common in other parts of its range.
- An **endangered species** is in danger of going extinct throughout its range, with the exception of insects pests that have been determined to present an overwhelming risk to man.

Are there any species that have been taken off the endangered list?

- Since 1973, only six species have been removed from the endangered species list in the U.S., including the American alligator and the gray whale.

- Elephant seals, sea otters, pelicans, eagles, ospreys, falcons, and some whales have been increasing in numbers, lessening their risk of extinction.
- However, another 1000 species have been added during the same time period.
- The net increase in listed species is due to the fact that habitat is being altered at an accelerating rate, that the recovery plans for endangered species have been underfunded, and the fact that new species are added as we learn more about their population sizes.

Don't species go extinct naturally?

- All species eventually will become extinct, but many alive today would not be expected to go extinct so soon.
- Approximately 99 % of all species that have ever existed have gone extinct.
- The average longevity of a species in the geological record is 10 million years.
- Prior to the Industrial Revolution, rates of extinction have been low (1 species extinction/year)
- Current rates of extinction are much higher than this (100 - 1000 species extinctions/year).
- During the "mass extinction" periods, the causes of which are uncertain but may be due to asteroid impacts or massive volcanic eruptions, 53 % of marine animal species went extinct (250 million years ago), and the dinosaurs went extinct (65 million years ago) (Figure 14.6).

What are the natural causes of extinction?

- After a population has become endangered because of various factors (See the *Frequently Asked Question*: **"How do humans cause local extinction and global extinction?"**), extinction can occur due to a number of factors:
  - **Population Risk** - Random variation in population birth rates and death rates can cause an already low population to decline to the point of extinction. This is especially true with species that have been reduced to a single population below 500 individuals.
  - **Environmental Risk** - Random variation in physical or biological environmental factors such as symbionts, food sources, predators, weather, can cause an already low population to decline to the point of extinction. For example, the heath hen went extinct because the population of goshawks, a predator, increased in the last remaining area where a preserve for the heath hen had been established.
  - **Natural Catastrophe** - Fires, floods, hurricanes, tornadoes, volcanoes, meteorite impacts are potentially capable of extirpating all individuals of a small population. If the population were the last one of a species, extinction would occur.
  - **Genetic Risk** - Genetic diversity within an already small population size is typically low. Changes in the genetic makeup of a population due to mutation or genetic drift (random loss of certain alleles from a population) can cause extinction (see Chapter 7).

How do humans cause local extinction and global extinction?

- Humans can cause species to become endangered, which can lead to extinction either locally or globally, in the following ways:
  - **Hunting and harvesting** - Humans prey upon animals and harvest plants to near extinction for sport, food, commercial sales of animal parts (fur, ivory tusks, etc.), or to control it as an unwanted pest species. The black rhinoceros and elephants are endangered because poachers hunt them for their ivory tusks, parrots are endangered as they are stolen from the jungle for use as pets, and passenger pigeons were driven to extinction by sport hunting.
  - **Habitat destruction** - Habitat for animals and plants is continually altered as humans encroach on wild areas. In a sense, as human "habitat" (cities, suburbs, farms, roads, shopping malls, etc.) expands, habitat for wild biota decreases. This is the major cause of species declines worldwide.
  - **Exotic species introduction** - Species from outside an area (exotic species) are introduced either intentionally or unintentionally and cause the extinction of native species. Examples of exotic species that are known to have caused extinctions of native species include the gypsy moth (intentional, for silk production), kudzu vine (intentional, for soil erosion control), walking catfish (intentional, for pets), the zebra muscle (unintentional in the ballast water of ships), Dutch Elm Disease (unintentional) and American chestnut blight (unintentional). Such common animals as cats, dogs, rats, and rabbits have driven to extinction or decimated thousands of endemic species, especially on islands.
  - **Pollution of the environment** - Pollution from pesticides and other toxic chemicals can biomagnify and produce developmental abnormalities in populations of biota. As a result, natural rates of reproduction of these species decline. Biomagnification in the food chain of DDT sprayed to control mosquitoes caused the decline of the bald eagle and other raptors in the late 1970s.

What is the difference between global extinction and local extinction?

- **Local extinction** is extinction of a population within a small area or region. For example, a species like the gray wolf may have once existed in your state, but is now locally extinct. The gray wolf still lives in Minnesota, Montana and Alaska.
- **Global extinction** is extinction from the Earth. This is the normal meaning of extinction.
- Local extinctions precede global extinctions. It is the accumulation of many local extinctions that leads to global extinction.

Are we conserving individual species or total biological diversity?

- We are interested in conserving both, but these are somewhat different goals.
- For example, the conservation of spotted owls requires that we preserve old growth forest (see Chapter 13) , but we may also want to preserve the greatest number of species as possible. This might require us to manage for early forest successional stages as well as old growth.

- Another example comes from the Everglades, where the Endangered Species Act prevents any alteration of an endangered species habitat. The Snail kite, an endangered hawk-like bird, nests in some areas that will be altered if water flow is restored in the Everglades to the natural levels. Overall biodiversity will be protected better if the natural flows are restored, but the Snail kite populations may suffer.

What is wildlife management?

- Wildlife resources are managed largely by managing the habitat and harvest rates on wildlife populations.
- The goal of wildlife management is to maintain sustainable harvests of wild animals from natural habitats.
- In the past, the harvests were done mostly for sport or subsistence hunting.
- Recently, a management goal of "game ranching" has been added in some areas. "Game ranching" is a practice of allowing wild species like zebras and impalas to grow on natural habitat, then harvesting commercially.

What is a fishery?

- A fishery is a predator-prey interaction in which humans are the predators and the prey are aquatic organisms. Thus, humans must always be part of a fishery management plan.
- The aquatic organisms are typically fishes of various species, but may also be invertebrates (like clams, oysters, shrimp, and lobsters) and also vertebrates (like turtles, seals, and whales).
- A **commercial fishery** is one in which the sole purpose is to capture aquatic organisms for profit.
- A **recreational fishery** is one in which the harvest of aquatic species is for sport.
- A **subsistence fishery** is one in which harvest is for food for the fishers.

How much fishery production is there worldwide?

- The annual wild harvest for all world fisheries is more than 101.4 million metric tons per year (130 million U.S. tons/year).
- Of this amount, marine fisheries produce 82 million metric tons per year (90.5 million U.S. tons/year), with the balance coming from fresh water fisheries.
- The increase in fishery production has declined to 2 % per year from the rapid rate of increase (5 % per year) that was apparent in the post war period (1950-1970).

What is a fish stock?

- A fish stock is simply a population.

What is overfishing?

- When more fish are harvested than can be replaced via natural growth and reproduction. Basically, this is non-sustainable harvest of a fish stock.

What fisheries have been overfished?

- There have been many notable cases, some of which have occurred even with active fishery management (Table 13.3).
- The most famous of these was the Peruvian anchovy, which reached a peak in 1970 at 10 million metric tons, then has been in decline ever since.
- Other overfished stocks include the North Atlantic cod and haddock off the New England coast, the Atlantic herring, the Atlantic menhaden, the North Sea herring, the Arctonorwegian cod, the Pacific salmon, and the Pacific sardines. It is also possible to overfish invertebrates like oysters in the Chesapeake Bay in the U.S.

What happened in the case of the Peruvian anchovy fishery collapse?

- This fishery has as its base the great wind-driven upwelling zone off the Peruvian coast, which brings nutrient-laden cool water from depth to the surface.
- This upwelling stimulates an annual phytoplankton bloom, which is transferred up the food web to zooplankton and then the anchovies (this is a very short food chain).
- The anchovies were harvested in order to make animal food, largely.
- The maximum rates of harvest occurred in 1970 when 10 million metric tons were landed (or about a tenth of the world's total fish catch) , but in just 2 years, the catch had dropped to 1.8 million metric tons. Two things led to this change: overfishing and El Nino.
- The **El Nino** is a condition in which the surface winds shift, the upwelling stops, and the phytoplankton-based food chain is transferring a much smaller amount of energy to the anchovies. This El Nino is part of a climatic pattern that is unpredictable, but happens every 8 - 12 years.
- However, after the El Nino year, the anchovies have remained rare. It appears that a combination of heavy fishing pressure along with the unpredictable cessation of upwelling , which caused the food base to decline, caused a long-term decline in anchovy abundance.

How are fisheries managed?

- Even fisheries under science-based management plans have collapsed; so the management has been poor in the past.
- Current management models rely on the logistic growth curve and mathematical models based on the concept of maximum sustainable yield (MSY; see Chapter 12). The biomass of fishes present in a region of the ocean and the carrying capacity (K) for that species is estimated by biologists using fishery catch data. Then, a portion of those fishes (usually, an amount that will lower the population to a size of K/2) are allowed to be harvested and this amount is allocated among commercial, recreational, and subsistence fishers according to guidelines established by regional fishery management councils.
- In the U.S., the regional councils are set up under the **Magnuson Fishery Conservation Act** and they govern fisheries in federal waters (3 miles - 200 miles offshore), with individual state management councils assuming management in waters closer than 3 miles.
- Some species have small stocks and slow population growth rates (tuna and sharks) and thus low amounts of allowable catch (called quotas) are maintained. Other

species have much larger populations and there are essentially no limits on how many can be harvested, but there may be limits on where or when they are harvested because of the unintentional catch, or **bycatch**, of other species (when shrimp are trawled, juvenile stages of other fishes, such as weakfish and red snapper, are caught in the nets).

- Generally, because of the uncertainty over how big the stock sizes actually are (biologists must rely on poor data sets with huge error estimates), there is great controversy over the size of the quota to set (fishermen always want larger quotas).
- Political arguments have developed routinely and invariably have led to overharvest as the upper end of the biologist estimates are used to set quotas.
- The fishermen win larger and larger quotas each year, until the stock collapses.

How many whales are there remaining in the world?

- There are nine species of commercially harvested whales that are considered threatened or endangered remaining in the world's oceans (Table 13.4).
- These whales were driven to near extinction largely due to the overharvest by whaling ships. The rarest of these whales are the right whale and the blue whale, of which less than 5000 remain in all the world's oceans.
- The blue whale is the world's largest animal, but has remained an endangered species despite the protection from whaling that it has received.
- At the other end of the spectrum, the Minke whale, which number 700,000 or more, is not considered endangered. This is a smaller whale for which there is a limited, but still controversial, hunt by native subsistence fishers in Arctic waters. This whale has made up some of the diet and is used traditionally by native peoples in Greenland, Tonga, Japan, and among the Eskimo people in Alaska and Canada.
- The gray whale, which was once considered endangered, has recently been removed from the endangered species list and is now regarded as threatened. There are now approximately 23,000 gray whales.

What did the Marine Mammal Protection Act (1972) do?

- This legislation passed by the U.S. Congress gave special protected status to whales, seals, dolphins, sea otters and other marine mammals.
- Although many of these animals are endangered and have very low population sizes, some of them are far from endangered and appear to be undergoing population booms. Nonetheless, it is illegal to harm or kill these animals in any way.
- Fishermen who inadvertently kill marine mammals (a bycatch problem) in their fishing operation can be subject to fines. This was a major problem in the tuna purse seine fishery of the Eastern Pacific Ocean, where over 100,000 spinner, spotted, and common dolphins were accidentally caught in the fishing nets because they feed along with the yellowfin tuna in schools.

What is the logistic growth curve?

- This concept was originally introduced in Chapter 5.
- The logistic growth curve is the S-shaped curve typically used in mathematical models of population growth. A plot of this curve shows **time (t)** in years (for most

populations) on the x-axis versus number of organisms in the population at each time period **(N)** on the y-axis. (Figure 14.8).

- The population following a logistic growth curve starts out growing nearly at an exponential rate, but then gradually slows due to competition for food and shelter resources among individuals in the population as the **logistic carrying capacity (K)** of the environment is reached. This is the maximum number of organisms that can be maintained indefinitely in that environment if the population meets the assumptions of the logistic model. K is the asymptote of such logistic plots.
- The logistic curve has been widely used in modeling populations of fishes, trees, wildlife, and even humans. Demographers use the logistic curve to predict K or the maximum number of organisms that will be in an area.
- However, very few organisms follow the curve exactly, and the assumptions made by scientists in order to use the model are rarely achieved in nature.
- The logistic model assumes that the population size at any time period can be determined *solely* by the density of the population at the previous time period; this is unlikely to be true in all cases.
- It is rare that populations stabilize at a single value of K, but rather they fluctuate as weather, interactions with other species, and random factors in ecosystems change the carrying capacity.

What is the Maximum Sustainable Yield?

- Maximum Sustainable Yield (MSY) is a mathematical model based on the logistic equation for a population; K is estimated from population growth data (Figure 14.8a).
- Harvest limits are determined by assuming that the population is growing at its maximum rate at one half of the carrying capacity or K/2, and using the growth rate of the population at K/2 as the harvest amount.
- Another way to visualize the MSY model and the underlying logistic is to plot the population growth rate or the **potential sustainable harvest** (also called the surplus of recruitment over loss, yield per recruit, annual percentage increase in population) versus population size (N) (Figure 14.8b). The population size is x-axis in this figure, but it was the y-axis in Figure 14.8a. This curve shows a low potential harvest at small and large values of N, i.e, when the population is large or small. This is when the population growth rate is minimal. At intermediate sizes of N, specifically at N= K/2, the population has the greatest potential sustainable harvest. The potential sustainable harvest at N=K/2 is called the **maximum sustainable yield**.

Is the MSY model appropriate for managing wildlife and fisheries?

- The traditional view of wildlife and fishery managers was that the assumptions of the MSY model were largely met, and that the model was an appropriate choice for setting sustainable harvest rates.
- Many managers are coming to grips with the reality that natural populations do not follow the logistic growth curve, that populations are affected by changes in the physical and biotic environment as well as the harvest rate and the number of individuals in the population at the previous time interval.

- These realities of fluctuating populations make the managers' jobs much more difficult, because the exact size of the population cannot be predicted, so harvest rates must be set conservatively. If the population is actually smaller than the logistic would predict, then the population could be overharvested, because the MSY would allow for too large a harvest.
- Many natural resource management failures, especially in fisheries with fish populations that are highly variable due to climatic variations, can be attributed to the exclusive reliance on MSY models.
- A better approach involves using Optimal Sustainable Yield models.

What is the Optimal Sustainable Yield and how does it differ from Maximum Sustainable Yield?

- Optimum Sustainable Yield (OSY) is a model for determining a harvest level that considers other factors besides the logistic growth curve and the maximum sustainable yield (MSY).
- OSY takes into consideration factors such as environmental variability of the ecosystem and effects on other species in the ecosystem.
- For example, OSY was used by the International Whaling Commission as a way of maximizing the population size rather than the harvest rate for whales. OSY has determined that harvest rates should be zero for a while for endangered species of whales. An MSY model would have predicted at least some amount of harvest could be done, even though it might have been just a few whales.

What is the minimum viable population size?

- The minimum viable population size is the smallest number of individuals that have a reasonable chance of persisting.
- A population size of 50 breeding individuals must be maintained in order to avoid inbreeding and genetic drift (random loss of genetic material) problems.
- Because most individuals do not breed (they are too young or too old), a safety factor is added to this minimum. A population should be at least 500 breeding individuals or 10 times the number of breeders that is needed to avoid genetic problems.
- Nature preserve areas are influenced by this figure. Because large animals need more habitat space than small animals to obtain their food, nature preserves must reflect the minimum population size multiplied by the home range for the largest species. For example, wolves need 26 km^2 each to find food, so a minimum viable population would need 13,000 km^2, or 5,000 mi^2 , of nature preserve to maintain the 500 wolves. If a nature preserve that is large enough cannot be set aside, then the 500 population size may be maintained artificially by moving wolves into other populations periodically for breeding purposes.

Ecology In Your Backyard

- What are the endangered species in your state? Every state has some endangered animals or plants that either are still present or have populations that have been driven to local extinction. Find out what species are still left in your area and see if you can help them by contributing money or time to a local conservation group or agency.
- Contribute to the non-game wildlife fund for your state. These funds are used to purchase land or restore habitat for endangered species. You can often check a box on your income tax or buy a special license plate to contribute to these funds.
- Join an environmental group like *The Nature Conservancy*. This group maintains a low profile, but is very effective at preventing the loss of critical habitats for endangered species. This group is made up of professional ecologists who are committed to protecting endangered species using nature preserves. Contributions are used to purchase land and keep it from being developed. Check out their website in the Ecolinks section below.
- Do you own a pet? If you do, do you know where it came from? Was it taken from the wild (as many birds, fish, reptiles, and other exotic species are) or was it bred in captivity? Most likely, if you have a dog or a cat, they were reared in captivity. If your pet is from the wild, do you know what the population status of your pet is? Find out if the species is native or exotic, and don't release it if it's an exotic (This is especially true for fish and reptiles!). Is it a threatened or endangered species? See if you can find out below on the Ecolinks section in one of the endangered species lists.
- Do you eat fish or shellfish?
- Where does the fish and shellfish you eat come from?
- What methods of harvest were used to capture the fish or shellfish?
- Is there a bycatch problem in the fishery that your seafood comes from?
- What species are being caught incidentally?

Links In The Library

- Bardach, J. E., J. H. Ryther, and W. O. McLarney. 1972. *Aquaculture: The Farming and Husbandry of Freshwater and Marine Organisms*. Wiley Interscience, New York, NY. 868 pp.
- Bell, F.W. 1978. *Food From the Sea: The Economics and Politics of Ocean Fisheries*. Westview Press. 380 pp.
- Erwin, D. H. 1966. The mother of mass extinctions. Scientific American, July 1996, 275(1): 72-78.

- Fordham, S.V. 1996. *New England Groundfish: From Glory to Grief. A Portrait of America's Most Devastated Fishery*. Center for Marine Conservation. Washington, DC.
- Robinson, S.K. 1996. Nest losses, nest gains. Natural History Magazine, July 1996, 105 (7): 40-47.
- Robinson, S. K. 1997. The case of the missing songbirds. Consequences 3 (1): 2 - 15.
- Sherman, K., L. M. Alexander, and B. D. Gold (editors). 1990. *Large Marine Ecosystems: Patterns, Processes, and Yields*. American Association for the Advancement of Science. 242 pp.
- Terborgh, J. 1992. Why American songbirds are vanishing. Scientific American, May 1992.
- Terborgh, J. 1989. *Where Have All the Songbirds Gone?* Princeton University Press, Princeton, NJ 207 pp.
- Ward, P. 1994. *The End of Evolution: A Journey in Search of Clues to the Third Mass Extinction Facing Planet Earth*. Bantam Books, New York, 301 pp.
- Warner, W.W. 1994. *Beautiful Swimmers; Watermen, Crabs, and the Chesapeake Bay*. Little, Brown, and Company, Boston.

Ecotest

1. The Endangered Species Act authorizes the U.S. Fish and Wildlife Service to prepare a _________________ for each listed species.
a. recovery plan
b. list of known property owners that have the species on their land
c. wildlife preserve
d. captive breeding program like the one used for California Condors

2. A population of which of the following animals has the best chance of surviving in a small (< 10 ha) ecological reserve of appropriate habitat (i.e., one in which that species is normally found)?
a. Bengal tiger
b. Spotted owl
c. Gray wolf
d. White-footed mouse

3. Since the Endangered Species Act was passed in the United States, the number of listed species has:
a. decreased as conservation measures have improved.
b. decreased because of the extinction of many listed species.
c. decreased because of political pressures.

d. increased as new information about animal population dynamics has become available.

4. Since the American alligator was listed as an endangered species in 1967, its population has:
a. decreased to the point where extinction is inevitable.
b. remained about the same.
c. increased somewhat but is still in danger of extinction.
d. increased dramatically and is no longer considered endangered.

5. Many people enjoy simply observing, identifying, and learning about wildlife (nondomestic) species of plants and animals. In addition to these aesthetic uses, wildlife provide other benefits to humans, such as:
a. clothing from skins and furs.
b. drugs and medical products from wild plants.
c. food from sport hunting, fishing, and plant gathering.
d. All of these choices are true.

6. When plants and animals are introduced into areas that are completely new to them, they:
a. often become destructive pests.
b. often outcompete native animals.
c. often increase because they have few natural predators.
d. All of these choices are true.

7. In the course of evolutionary history,
a. humans have caused most extinctions.
b. about one-half of all species that have ever existed have gone extinct.
c. extinction has only occurred in conjunction with meteor impacts.
d. about 99 % of all species that have ever existed have gone extinct.
e. None of these are correct.

8. A utilitarian justification for protection of biological diversity in a tropical rainforest is:
a. a high diversity of birds will attract birdwatchers.
b. medicinal products may be derived from plants in the forest.
c. food items that may be grown for human consumption will be found.
d. climate conditions worldwide will be maintained if deforestation is stopped.
e. All of these are utilitarian justifications.

9. Populations following a logistic growth curve:
a. show exponential growth at low population sizes when resources are unlimited.
b. slow gradually as competition for resources increases.
c. reach an asymptote when the carrying capacity of the system is reached.
d. All of these are correct.

10. The number of individuals (N) at which a population has the greatest potential sustainable harvest is called:
a. maximum economic value.
b. carrying capacity.
c. maximum sustainable yield.
d. optimal growth.

11. Humans can cause extinctions as a result of:
a. hunting and harvesting.
b. habitat destruction.
c. introducing exotic species.
d. pollution of the environment.
e. All of these are correct.

12. The random loss of genetic material from a population is defined as:
a. eugenics.
b. genetic drift.
c. genetic manipulation.
d. natural selection.

13. The cause of the decline in whale populations has been attributed to large-scale:
a. pollution.
b. reduction in whale food supplies from overfishing.
c. whaling.
d. habitat destruction.
e. All of these are correct.

14. A fishery is a:
a. complex predator-prey relationship between human and aquatic animals.
b. a marketplace where fishes are bought and sold.
c. an environment in which many diverse fish species are found.

15. In the logistic mathematical model used to estimate fish populations and manage them, how are the harvest limits under MSY (maximum sustainable yield) determined?
a. Harvest amount = Carrying capacity (K) = MSY
b. MSY harvest = K/10
c. MSY harvest = an amount that will lower the population to K/2
d. MSY= r*N[(K-N)/K]
e. None of these are correct.

Chapter 15 - ENVIRONMENTAL HEALTH, POLLUTION AND TOXICOLOGY

The Big Picture

Pollutants are emitted from a great variety of natural sources, from volcanoes that alter entire landscapes with deposits of ash and expose plants and animals to toxic vapors, to photosynthesis itself which produces oxygen, a pollutant to anaerobic microbes. Species evolved tolerances in response to pollutants. In recent decades (especially during the second half of the twentieth century) hundreds of thousands of new, synthetic pollutants have been introduced into the environment. Some of these pollutants are highly toxic and persist in the environment and in biological tissue. As a result species populations that are live in areas receiving pollutants in high concentrations or species that have low tolerances for pollutants may experience detrimental effects at the organism and population scale. The effects of pollutants on human health are well founded. Considerable research (at great expense) has been conducted to determine how pollutants affect humans and how exposure to pollutants can be minimized. Risk assessment is necessary to objectively ascertain the risks to human and environmental health posed by specific pollutants. Certain risks to human health are well documented and well known (such as cancer risks due to smoking tobacco), other risks are poorly understood and subject to misinformation or misinterpretation (such as cancer risks associated with asbestos).

Frequently Asked Questions

What are the differences between pollution, contamination, toxic, and carcinogenic?

- **Pollution** is the state of being impure, defiled, dirty, or unclean.
- **Contamination** is pollution that makes something unfit for a particular use by the introduction of undesirable materials.
- **Toxic** refers to pollution by poisonous substances.
- **Carcinogenic refers** to pollutants that have been determined to cause cancer.

How do organisms (including humans) respond to pollutants?

- Reactions to pollutants are highly variable.
- **Individuals vary in their response**; differences may be due to such factors as body size, age, general health, immune response, previous exposures, sensitivity, or other known or unknown causes.
- **Pollutants have a threshold**; below this threshold, exposure causes no effects; above the threshold, effects may be manifest.
- **Some responses to exposure are reversible**; other responses are not permanent.

- **Pollutants may be changed** by ecological processes and biological processes, thus their effects may be altered by these processes.
- The effects of multiple exposures to one or several pollutants may have **a synergetic effect**, where the combined effect of multiple exposures is greater than any single exposure.

How are pollutants introduced into the environment?

- Pollutant pathways are innumerable; some of these pathways are discussed in more detail in the chapters on water and air pollution (Chapters 21 nd 23), nuclear energy (Chapter 19), and agriculture (Chapter 12).
- Generally, however, pollutant sources are categorized as **point sources** or **area sources** (non-point sources).
- Point sources usually emit pollutants in high concentrations; point sources include wastewater discharge pipes, smokestacks, and accidental spills. Most point sources are more easily monitored, regulated, and remediated than area sources.
- Area sources are more diffuse than point sources. Concentrations from area sources are usually less concentrated than point sources, but they are also more difficult to detect, monitor, and control. Agricultural and urban runoff, and automobile exhaust are common examples of area sources.

How is the amount of pollution measured?

- In the United States, the EPA has standard methods (published guidelines) that are used to collect and analyze samples of air, water, soil, and biological tissue for pollutants. Primarily for legal purposes, the EPA standard methods must be followed by agencies or private concerns that are collecting, handling, or analyzing samples for pollution analysis. Analytical laboratories conducting pollutant analysis must be EPA-approved.
- The EPA publishes a list of pollutants along with the threshold concentrations for human health and the environment. Most threshold levels have been determined in laboratory settings using non-human test subjects and clinical studies; as indicated earlier, the actual effects for individuals are highly variable.
- The units used to measure and report pollutant concentrations depend upon the questions being asked. For example, if you want to know the total amount of cadmium (a potentially toxic heavy metal) discharged in a watershed, the appropriate units may be tons discharged per year. If you are interested in the threshold concentration for human health, the appropriate units for cadmium would be μg/L.
- Commonly reported units include:
    - metric tons/year for watershed or air shed discharges or emissions
    - m^3 per day or year or gallons per day or year for wastewater discharges
    - ppb for extremely toxic substances; or ppm for highly toxic substances; or percent for less toxic substances
    - μg/L or mg/L are standard units for water pollutants
    - microorganisms or colonies per mL for waterborne pathogens

What are infectious agents?

- Infectious diseases that are transmitted via the environment (water, air, soil, contaminated food) comprise a serious risk to human health.
- Environmentally transmitted infectious diseases, usually a consequence of poor sanitation, are the greatest cause of death in developing countries.
- In the U.S. and other more developed countries, mortality from infectious diseases is less likely but serious illness occurs none-the-less; Legionellosis, Giardiasis, Salmomella, malaria, Lyme Borreliosis, and cryptosporidosis are but a few of these diseases.

What are toxic heavy metals?

- Naturally occurring metallic elements - mercury, lead, cadmium, nickel, platinum, bismuth, arsenic, selenium, vanadium, chromium, thallium, even gold and silver - can be toxic in high enough concentrations.
- Heavy metals do not degrade in the environment and are especially persistent in biological tissue.
- Heavy metals originate from natural sources (rock and soil) but are concentrated and mobilized by human activities. Obviously many metals are mined, the ores refined, and the metal products are used in industry and manufactured products. Metals may also be inadvertently concentrated and mobilized by land use activities that disrupt soils (forestry, agriculture, and construction). Metals may be contained in emissions from smoke stacks and discharges from waste water discharge pipes.
- The physiological effects of heavy metals are commonly manifest in tissue of the nervous system and other internal organs. The effects are especially pronounced in fetuses and children.
- Because heavy metals persist in biological tissue, they are readily transferred and **biomagnified** in food chains.
- The total amount of heavy metals in an organism is called the **body burden.**
- Lead, mercury, and selenium are three heavy metals whose health risks have been well publicized in recent decades. Lead contamination is problematic in old house paint and in lead water pipes, mercury contamination is present in many marine and freshwater fish (even fish that live far from sources of pollution), and selenium has caused physiological problems for wildlife species in California.

What organic compounds are pollutants?

- Organic compounds are produced naturally (the biota is composed of organic molecules) or produced synthetically.
- More than 4 million organic compounds are listed by the American Chemical Society. Synthetic organic molecules are used in industrial processing and consumer products, pesticides, pharmaceuticals, and food additives.
- Some **natural organic compounds** are toxic; for example the alkaloids in poison ivy or nicotine in cigarettes.
- Some **synthetic organic compounds** are toxic, such as the chlorinated hydrocarbons (e.g., PCB, DDT, and 2,4,5T, chlordane, Agent Orange); the organophosphates (e.g., malathion and parathion); and the carbamates (e.g., carbaryl or Sevin Dust, methiocarb).

- DDT, PCB, and dioxin are three types of synthetic compounds that have received national and international attention.
- **DDT**, an insecticide, was one of the main topics discussed in *Silent Spring* by Rachel Carson; following the publication of this book, government, industry, and citizens began reevaluating chemical usage and their effects on human and environmental health.
- **PCB**s are widely used in electrical transformers. Inadvertent or illegal disposal of PCBs have contaminated fish in lakes and rivers, such as the Hudson River in New York.
- **Dioxin** is actually a suite of toxic chlorinated hydrocarbons that may be found in herbicides (Agent Orange) and pulp and paper mill wastewater. Times Beach, Missouri is an infamous site of dioxin contamination.

What is radiation pollution?
- Intensive or prolonged exposure to ultra high frequency radiation from radioisotopes can cause damage to cellular DNA. Certain types of birth defects and cancers have been attributed to exposure to radiation.
- Radioactive isotopes occur naturally in rock formations and in some locations may emit sufficient radiation to present a problem to human health.
- The greatest risk to most individuals is from unnatural sources; primarily, medical X-rays and radioactive wastes or emission from nuclear power plants.
- Radiation risks are discussed in more detail in Chapter 18.

What is thermal pollution?
- Thermal pollution is the abnormal heating or cooling of natural waters.
- Thermal pollution does not usually present a direct threat to human health, but aquatic ecosystems can be significantly altered by temperature changes.
- Dissolved oxygen concentrations are dependent upon water; increased water temperature decreases dissolved oxygen concentration temperature, although other factors are involved as well. Abnormally warmed waters cannot support fish and other aquatic fish that require cool waters with high dissolved oxygen concentrations (Figure 14.6).
- Spawning in many aquatic species is dependent upon water temperature; abnormal heat or cooling of waters disrupts spawning behavior and success.
- Abnormal heat is attributed to warm water discharges from power plants (Figure 14.7), runoff from roads and parking lots, and the removal of streamside vegetation that would otherwise provide shade.
- Abnormal cooling occurs when cool or cold water is drawn from the bottom of hydropower lakes and reservoirs then discharged downstream.

What is particulate pollution?
- Particulates are small aerosol particles of dust.
- Particulates originate from natural sources such as volcanoes, wildfires, and dust storms.

- Particulates may also originate from anthropogenic sources or causes such as combustion of fossil fuels and biomass, dust from agricultural fields, and land use changes that exacerbate wind erosion.
- Most particulates are nontoxic but can contaminate lungs and cause respiratory and cardiovascular problems.
- Some particulates contain toxic materials.
- Particulates can interfere with photosynthesis in plants and lung function in animals.
- Chapter 23 addresses particulate pollutants.

Why is asbestos a pollutant?
- Asbestos is a natural mineral that occurs in two or more different types.
- **Chrysolite**-type asbestos accounts for about 95% of the usage in the United States. Usage is for pipe insulation, brake lining, and fire-retardant applications. Under normal conditions, this type of asbestos has not been found to be especially hazardous to human health.
- **Crocidolite**-type asbestos is less widely used but is a significant health risk. Lung damage and certain cancers have been attributed to exposure.
- Because most exposure is to chrysolite asbestos instead of crocidolite asbestos the health risks of asbestos exposure may be overstated, even though considerable time, effort, and expense have been devoted to asbestos removal from schools and other public buildings.

Why are electromagnetic fields a type of pollution?
- The human health risk associated with exposure to electromagnetic fields is difficult to measure and is controversial.
- Some epidemiological studies suggest there is a link between long-term exposure to electromagnetic fields and certain cancers, but the studies are not definitive.

Why is noise considered a pollutant?
- Noise is unwanted sound.
- Sound waves moving through the atmosphere are measured in units of decibels (dB), a logarithmic scale (Table 15.1).
- Noise at relatively low decibel levels may constitute an annoyance, thus stress and stress-related health problems may arise.
- Noise at higher decibel levels can cause physiological damage to ear structures resulting in partial or total hearing loss.
- Prolonged exposure to high levels of noise or short-term exposure to very high levels can permanently damage hearing.

What is voluntary exposure?
- The voluntary intake of tobacco products, alcohol, and the illegal use or misuse of drugs present a much greater health threat to more people than exposure to any other toxin.
- An estimated 30% of all cancers and 80% of all lung cancers in the U.S. are linked to smoking tobacco.

- Second-hand smoke is now recognized as a serious health threat to non-smokers.
- Many people in the U.S. have heeded the cautions against smoking and have quit or never started smoking. Others have not paid attention to the warnings and are continuing to risk their current and future health status by taking up smoking and other uses of tobacco.
- About 70% of all people in the U.S. drink alcohol; in moderation for most, but abused by others. Alcohol abuse is physically debilitating and is responsible for a great deal of domestic and societal ills.
- Illegal drug use continues to plague society. The addictive qualities of some drugs and detrimental health and societal effects of drug use are well documented. Illegal drug use in the U.S. supports a massive black-market trade that extends from many segments of our society to far beyond our national borders.

What are the general effects of pollutants?

- As discussed earlier, pollutant effects on individual humans are variable dependent upon the individual's physical characteristics, intensity or duration of exposure, and pollutant type.
- Many toxins target specific organs or organ systems (Figure 15.5). Others have more widespread effects. A general decline in health associated with a specific toxin may precipitate illness or disease not directly attributable to the toxin.
- Most toxicological studies have focused on human health problems. Most of the state and federal pollutant water and air quality standards are based on pollutant threshold levels from these human health studies.
- The responses of wildlife to pollutant exposure may be similar to some human responses, but because wildlife interrelates with the environment differently than humans, it is erroneous to assume the pollutant threshold levels that are suitable for humans are applicable to wildlife (Table 15.2). For example, low-level concentrations of pollutants in streams may pose no risk to humans wading or swimming in streams, but to filter feeding freshwater mussels in the streambed, low-level concentrations may be lethal.

What is dose response?

- The health effects of chemicals, be they toxic or nontoxic, depend upon dosage (amount of exposure).
- Chemicals have a range of health effects and the effects vary among individuals.
- The health effects of chemical dosage can be predicted and depicted graphically as **dose-response curves** (Figures 15.12, 15.13, and 15.14).
- The **general dose-response curve** indicates that large dosages of a chemical may be toxic, small dosages may have no effect, but intermediate dosages may have beneficial effects (e.g., trace elements needed in our diet).
- A **toxic dose-response curve** can be graphed for environmental pollutants (Figure 15.12). Because the health effects of exposure to toxic chemicals vary within a population, a standard means of expressing the toxicity of a chemical is necessary.
- Toxicity of a particular chemical is usually reported as LD-50 and/or ED-50.

- Tests for chemical toxicity are conducted on non-human subjects, thus the effects of a particular dosage in humans may not be the same.
- **LD-50** (lethal dose- 50%) is the dosage at which 50% of a test population dies.
- **ED-50** (effective dose - 50%) is the dosage at which 50% of the test population shows an effect.
- For many chemicals, the **threshold** concentration at which exposure causes health problems is not known.
- Ideally, if there are no known health benefits to exposure at low concentrations and exposure at high concentrations is detrimental, the best option is to avoid exposure altogether. This becomes problematic, for example, because industrial and domestic waste treatment technologies do not adequately remove all pollutants from discharged material. For many chemicals, using more sophisticated waste treatment technologies is not cost effective.

What is the difference between an acute effect and a chronic effect of exposure to pollutants?

- **Acute effects** occur immediately or soon after exposure to pollutants. Often the exposure is well in excess of threshold levels. Some, but not all, acute effects are reversible.
- **Chronic effects** result from long-term exposure to low concentrations of pollutants. The source of chronic effects may be difficult to determine and the effects may not be manifest until later in an organism's life span. Many chronic effects are non-reversible.

How do dose-responses differ along ecological gradients?

- Organisms nearest the source of pollutant contamination have the most severe and acute effects. Organisms far from the source are exposed to pollutants in diluted concentrations and would be more likely to experience chronic effects.
- The tolerance of organisms varies along ecological gradients.
- **Tolerance** is an indication of the ability of an organism to resist or withstand exposure.
- Tolerance is dependent upon:
  - **behavior** (i.e., learning to avoid exposure to pollutants),
  - **physiology** (i.e., physiological adjustment to a pollutant that increases tolerance; this includes **detoxification** of pollutants via metabolic process),
  - **genetic adaptability** (i.e., the survival of more resistant individuals in a population thus altering the gene pool; pesticide resistance in insects is an example).
- Aquatic invertebrate species are often used as an index of environmental quality in streams. The presence and abundance of invertebrate species changes along an ecological gradient from more severely polluted sites to less severely polluted sites.

What is risk assessment?

- Assessing risk is important if strategies are to be developed that will minimize human and environmental health problems that stem from exposure to toxins or other types of hazards.

- **Risk assessment** involves four steps:
  - **Identification of the hazard** (i.e., determining the effects of a particular hazard),
  - **Dose-response assessment** (i.e., determining the dosage at which health problems arise),
  - **Exposure assessment** (i.e., determining the intensity, duration, or frequency of exposure),
  - **Risk characterization** (i.e., determining the magnitude of the problem for human and environmental health).
- By evaluating these aspects of risk assessment, risk management decisions can be made that integrate technical, legal, political, social, and economic factors.

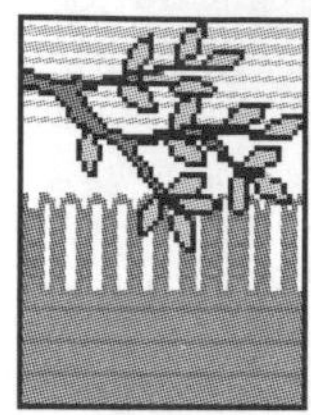

Ecology In Your Backyard

- What pollutants do you think present the greatest risk to your health or to ecosystems in your area? Remember, recognition of a potential threat is an important first step. Potential sources to consider include:
- Are you exposed to indoor air pollutants, especially tobacco smoke?
- Are you exposed to noise pollution at concerts?
- Are urban air advisories issued for your area? (Check the air quality index published in local newspapers to find out what pollutant groups present the greatest risk.)
- What residual pesticides might be found in the foods you eat?
- In rural areas, are you ever exposed to drifting pesticides from crop dusting?
- Have health advisories been issued against consuming fish from local lakes, rivers, and streams?
- Are there any toxic waste sites in your area? What is the status of discharge or cleanup?
- You might find answers to some of these questions by searching websites maintained by government agencies and independent organizations that provide information about pollutants and pollutant risks for locations throughout the United States.
- Check the "right to know" databases on the Internet to find out which pollutants are a health or environmental threat in your watershed, air shed, or locality.

Links in the Library

- Briggs, S.A. and the Rachel Carson Council. 1992. *Basic Guide to Pesticides: Their Characteristics and Hazards*. Taylor and Francis, Washington.

- Colborn, T. D. Dumanoski, and J. P. Meyers. 1996. *Our Stolen Future*. Penguin Group, New York. 306 pp.
- Klassen, C.D, M.O. Amdur, and J.O Doull. 1986. *Toxicology: The Basic Science of Poisons*. Macmillan Publishing Company, New York.
- Hallenbeck, W.H. and K.M. Cunningham Burns. 1985. *Pesticides and Human Health*. Springer-Verlag, New York.
- Rodricks, J. V. 1992. *Calculated Risk: The Toxicity and Human Health Risks of Chemicals in Our Environment*. Cambridge University Press, New York. 256 pp.
- Worthington, C.R. 1991. *The Pesticide Manual: A World Compendium, 9th ed.* British Crop Protection Council.

Ecotest

1. The level or concentration at which a pollutant causes health problems is called the pollutant ______.
a. dosage
b. threshold
c. risk point
d. toxicity point

2. Pollutant sources such as runoff from agricultural and urban areas, and automobile exhaust are examples of ______ sources.
a. mobile
b. point
c. detectable
d. area

3. A measure of the total amount of heavy metals in an organism is its _______.
a. toxic load
b. body burden
c. chelation total
d. accumulation index

4. ______ is the name of a group of chlorinated hydrocarbons found in herbicides (like Agent Orange) and in wastewater from pulp and paper mills. The community of Times Beach, Missouri was severely contaminated by this synthetic toxin.
a. DDT
b. PCB
c. Dioxin
d. Mirex

5. Exposure to _____-type asbestos is especially hazardous to human health; other types of asbestos are less problematic.
a. nonfiberous
b. crocidolite
c. chrysolite
d. silicofibric

6. An estimated ___ % of all lung cancers in the United States can be attributed to smoking tobacco.
a. 80
b. 75
c. 50
d. 20

7. LD-50 is:
a. the point at which a pollutant is diluted to 50% of its initial concentration.
b. the point at which a pollutant is detoxified by 50% of its initial concentration.
c. the point at which 50% of a test population dies after exposure to a pollutant.
d. the point at which 50% of a sampled natural population test positive for a toxin.

8. The high concentration of toxins in the tissue of animals that feed high on the food chain is a consequence of ________.
a. toxification
b. high ED-50
c. low tolerance
d. biomagnification

9. Long-term exposure to low level concentrations of pollutants typically cause _______ health effects.
a. carcinogenic
b. nonadaptive
c. chronic
d. low-dosage

10. Three of the four steps involved in risk assessment are listed below; which is not one of those steps?
a. identification of the hazard
b. dose-response assessment
c. exposure assessment
d. remediation

11. All of the following are examples of a point source except:
a. acid rain.
b. stormwater collection pipe.
c. hog waste lagoon.
d. factory smokestack.

12. Thermal pollution presents the most direct threat to:
a. humans.
b. fish.
c. birds.
d. terrestrial insects.

13. Which of the following is not a toxic heavy metal?
a. lead
b. mercury
c. dioxin
d. selenium
e. chromium

14. Noise can be considered a pollutant because:
a. it can constitute an annoyance that can lead to stress-related health problems
b. it can cause physiological damage at high-enough levels
c. it is the part of sound that is "dirty", "unclean", or "impure", or basically the sum of unwanted sound.
d. All of these are correct.

15. Acute effects occur:
a. immediately or soon after exposure to pollution.
b. due to exposure to pollutants at levels that are well below threshold levels.
c. are always permanent.
d. All of these are correct

Chapter 16 - ENERGY: SOME BASICS

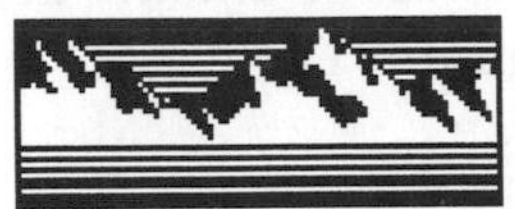

The Big Picture

Energy is a subject that is central to the understanding of ecology; it is also a commodity that we purchase every day. In fact, energy is central to what we do every day - work. Energy is the ability to do work. When you work, someone pays you for your expended energy. That money is used to buy food, which is itself energy in a different form, to replace the energy that you have burned in working. You also use your money to pay heating and electric bills, fill the gas tank of your car, or buy natural gas to cook your food. Thus, humans are part of the energy flow in the food web, and our monetary system is just a way of representing energy. These commercial sources of energy are derived from energy in various other parts of the ecosystem, including fossil fuels, wood, nuclear power, hydroelectric power, wind power, or geothermal power, but ultimately they all trace back to the sun's energy. In this chapter, the authors explain just what energy is, review the laws of thermodynamics, demonstrate how to calculate energy efficiency, describe various energy units, contrast energy with power, and show how energy consumption is calculated. In addition, they examine the alternative ways of meeting our future energy needs: the "hard path" of building more energy generating capacity and the "soft path" of energy conservation. The need to reassess national energy policy was realized during the summer of 2001 when the state of California experienced a energy crisis. Inappropriate energy choices coupled with an economic downturn triggered short-term energy shortages and "rolling blackouts". Unfortunately the situation in California may be a harbinger of future energy supply problems because of our continued reliance on unsecured and finite energy reserves and resources.

Frequently Asked Questions

Are energy crisis new?

- No, historically, societies have had to search for new energy supplies and/or new energy technologies as traditional supplies diminish.

What is energy?

- Energy is the ability to do work.
- **Work** is exerting a force over a distance (Work = force x distance).

What are the 1st & 2nd laws of thermodynamics? (Review Chapters 3 and 9)

- 1st Law of Thermodynamics states that energy may be converted from one form to another, but it is never lost from the system.

- 2nd Law of Thermodynamics states that during conversion from one type of energy to another, some energy is converted to heat, and the entropy of the system must increase.
- Heat and entropy are different. The Second Law says that there is always waste heat created. All of the energy in a system is not usable to do work. Entropy is a measure of the amount of unusable energy in a system. (Its units are not energy but energy/temperature.) The waste heat given off increases the amount of unusable energy increasing the entropy.

What are the various forms of energy?

- ***Potential Energy*** - This is stored energy (high quality energy) and can be used to do work.
    - It can be stored as:
    - Chemical energy - coal, oil, natural gas (fossil fuels)
    - Nuclear energy - stored in atomic bonds
    - Gravitational energy - water stored behind dams or a vehicle at the top of a hill
- ***Solar Energy*** - Light energy from sun
- ***Kinetic Energy*** - Kinetic energy is energy that an object has due to its state of motion.
- ***Heat*** - Energy that flows from a high temperature object to a low temperature object in a system; it can be used for work (low quality energy).

What happens when work is done?

- When work is done or an object is moved, the following sequence normally occurs:
- Potential energy --> Kinetic energy --> Heat
- Low entropy in system ----> High entropy in system
- High quality energy -----> Low quality energy
- Organized (ordered) energy ----> Disorganized (disordered) energy
- It is possible to work backwards here, for heat to cause motion of things and thus create kinetic energy and then potential energy; however when this occurs entropy of the entire system increases at each conversion.

What is entropy?

- Entropy is a measure of the amount of energy unavailable to do work in the system. It is the randomness of a system.
- As the molecules of the compounds in which energy is stored are burned, the bonds between the atoms are broken, releasing energy and making the distribution of the molecule's atoms more random.
- All the energy in a system eventually ends up in an unusable form. This means that entropy increases.
- Entropy is **NOT** energy. Its units are not energy but energy/temperature. When all of the energy is heat energy, the system is at the same temperature, and none of the energy is usable. Entropy is maximized.
- This is a one-way path; you can't go back to high quality energy from low quality energy.
- This is why energy cannot be recycled, and why new high-quality energy from the sun is always needed for the Earth's ecosystems.

What is friction?

- Friction is the force that converts kinetic energy into heat energy when work is done. It changes organized motion into disorganized motion.

What is energy efficiency?

- There are two kinds of energy efficiency: first-law and second-law efficiency.
- **First-law efficiency** is the ratio of total energy output (or the work done by the device) to total energy input (the amount of energy used to do work).
- **Second-law efficiency** is a ratio equal to the minimum amount of energy needed to perform a task divided by the actual energy used to perform the task. The minimum amount of energy is the input energy required under ideal (Carnot efficiency) conditions. Carnot efficiency is the maximum efficiency attainable for a given set of conditions (This type of ideal efficiency was first described by the French engineer Sadi Carnot in 1824).
- Second-law efficiency takes into account the fact that not all energy is available to do work (2nd Law of Thermodynamics).
- Second law efficiencies can often be improved because our actual machines are not ideal Carnot efficiency machines. For example, automobile engines can be redesigned so that we can get more MPG, but still go the same distance at the same speeds. The more closely we approach the ideal efficiency, the more energy we can save.
- An electrical power plant is 33 % efficient in terms of second-law efficiency. That means that for every 3 units of coal burned, 2 units are lost as heat and 1 unit is converted to electricity.

What is a heat engine?

- A heat engine produces work from heat.
- A coal-burning power plant is a heat engine, because it burns coal to heat water which becomes steam and turns a turbine that generates electricity.
- Many energy demanding technologies use heat engines to power them, including automobiles and steam engines.
- Heat engines produce a great deal of waste heat, which causes thermal pollution and contributes to global warming .
- Most objects that do work are actually heat engines. One important example is living organisms. They use heat energy released from their food to do work. To see a good example of heat energy in food, burn a potato chip (not the baked kind).

What is the energy consumption in United States?

- In 1999, the total energy consumption in the U.S. exceeded 95 exajoules. (1 exajoule is 10^{18} Joules, or approximately 1 quadrillion BTUs)
- The U.S. uses 25 % of the world's energy with 5 % of the world's population.
- Most of the U.S.'s energy comes from the fossil fuels (coal = 23 exajoules, natural gas = 25 exajoules, and oil = 33 exajoules) Ten exajoules come from other energy sources (hydroelectric, solar, wind).
- The overall first-law energy efficiency for the U.S. is 50 %. This means that half of all the energy we use to do work is lost into atmosphere as heat (42.5 exajoules in 1999). This heat loss to the atmosphere is part of the reason why the Earth is gradually warming.

- The residential/commercial and industrial sectors each consumed 36% of the total energy used in the U.S., the transportation sector consumed the remaining 26%.

What are the units used to measure energy?

- Energy is measured using the following units:
  - 1 joule = 9.481×10^{-4} BTU's = 0.239 cal = force of 1 Newton applied over 1 meter
  - 1 calorie = amount of heat needed to raise 1 g (or ml) of H_2O 1 °C = 4.18 joules
  - 1 Kilocalorie (or Kcal) = 1000 calories (dieters count Kcals, but call them "Calories")
  - 1 British Thermal Unit (BTU) = 252 calories = 1055 joules
  - 1 quad = 1 quadrillion BTU's = approximately 1 exajoule
  - 1 exajoule = 10^{18} joules or a billion gigajoules = approximately 1 quad = (9.481×10^{15}) BTU's
  - 1 kilowatt-hour = 3,600,000 joules

What is power and how does it differ from energy?

- Power is the rate of energy use (Energy used/unit time).
- Power is measured using the following units:
  - 1 watt = 1 joule/second = 3.413 BTU/hr = 4.34 cal/hr
  - 1 kilowatt = 1000 watts
  - 1 megawatt = 1000 kilowatts
  - 1 gigawatt = 1000 megawatts
  - 1 horsepower = 7.457×10^2 watts

What is cogeneration?

- This is a process of using waste heat from large power generating facilities to do useful work.
- For example, the location of a business building next door to a power-generating facility can save on heating costs. This is because the 2 units of heat produced for every 1 unit of electricity produced can be used to heat the building. This saves the business from having to reconvert electricity back into heat, as is frequently done (think of an electric range, electric water heaters, and electric home heat).
- Cogeneration represents an improvement in the overall efficiency of burning fossil fuels, so it creates less thermal pollution and CO_2 release.

What is fossil fuel?

- Coal, petroleum, and natural gas are fossil fuels. Through geologic processes, these fuels were derived from the remains of ancient organisms.
- Fossil fuels are of finite supply (see Chapter 17)

What are alternative energy sources?

- Solar, hydropower, hydrogen fuel cells, windpower, geothermal and tidal power are examples of alternative energy sources (see Chapter 18) Generally, alternative energy sources have less detrimental environmental impacts than fossil fuel sources.
- Nuclear power is an alternative to fossil fuel but the potential environmental consequences may be severe (see Chapter 19).

What types of energy consumption in the U.S. can be made more efficient?

- The three areas that have the greatest potential for energy savings are building design, industrial energy use, and automobiles.
- These three energy uses account for 60 % of the current consumption, so efficiency improvements will have the biggest impacts here.
- **Building design** - Use more passive solar (See Chapter 18), build with windows facing south, overhangs, use good insulation, double pane glass, etc.
- **Industrial energy use** - In the 1990s, industrial output has climbed while energy use by industry has stabilized; this means that more goods are produced per unit of energy today than in the 1970s. Industry has become more second-law efficient since the 1970s, due to the use of cogeneration facilities, solar energy, more efficient pumps and motors.
- **Automobile design** - In 1970, the average car got 14 mpg; today the average is 28 mpg (highway driving in both cases). Cars have become more fuel-efficient, without impairing anyone's ability to drive them around. Hybrid cars (gasoline-electric) are now available and achieve mileage ratings of up to 90 mpg (highway).

What is meant by the Hard Path and the Soft Path for energy policy?

- There are two choices ahead for energy policy in the U.S.: The hard path and soft path.
- The **hard path** means building more power generating capacity to meet an increasing demand for energy by a growing population in the U.S.. It is "hard" because it means increasing the "hardware" needed to make power.
- The **soft path** means becoming more energy efficient than we have been. By encouraging conservation and increasing energy efficiency, we can make more goods and provide more services with the same amount of energy. This path is "soft" because it means not building more power plants, but changing the behavior of people and the energy efficiency of machines.

What is integrated energy management?

- Integrated energy management increases reliance on a diversity of fuel types instead of one or few fuels.
- Fossil fuels are a part of integrated energy management but by using a diversity of fuels fossil fuel supplies can be extended as we move toward energy sustainability.

Ecology in Your Backyard

- How can you improve the energy efficiency in your life? Try some of these ideas:
    - Do you drive a car when you could ride a bike or walk instead? Both driving and biking produce CO_2, but driving contributes more CO_2 and from a different source (fossil fuels) than biking or walking. (When biking, the CO_2 comes from the food you have eaten, but that carbon came from the atmosphere very recently; the car's

CO_2 came from fossil organisms, so it is being reintroduced into the atmosphere after 300 million years of storage in the Earth).

- How is your water heated? Most homes and apartments have electric water heaters, which are wasteful because water must be heated twice: first at the power plant to make steam to turn the generators, then later to make the water hot at your house. A better idea is to use natural gas or solar. Solar is by far the best choice, but installation of a solar hot water heater can cost $5,000. This amount of investment can be recovered in 10 years of fuel bill savings, however.
- What kind of light bulbs do you use? If you use the traditional incandescent bulbs, you are wasting money and contributing to global warming as well. Compare the energy efficiency of fluorescent and incandescent bulbs in Table 16.1 in the text. If everyone switched to compact fluorescent bulbs from incandescent, the amount of energy used and CO_2 released would both decrease. Again, compact fluorescent bulbs cost more initially ($7.00 or more per bulb), but they last much longer and produce the same amount of light as an incandescent, with a lot less energy used.

Links In The Library

- Fickett, A.P. 1990. *Efficient use of energy.* Scientific American 263 (3): 157-163.
- Lovins, Amory B. 1990. *The negawatt revolution.* Across the Board, September 1990, pp. 18-23.
- Nixon, Will. 1991. *Energy for the next century.* E Magazine, May/June 1991, pp. 31-39.
- Perlin, J. 1991 . A Forest Journey : The Role of Wood in the Development of Civilization. Harvard University Press. Cambridge, Massachusetts.
- White, David C., Clinton J. Andrews, and Nancy W. Stauffer. 1992. *The new team: Electricity sources without carbon dioxide.* Technology Review, January 1992, pp. 42-50.

Ecotest

1. Sixty percent of the energy used in the United States is for:
a. industrial processes, home heating and air conditioning, and automobiles
b. aircraft, automobiles, shipping and trains
c. military and government uses
d.agriculture and industrial processing

2. One proven way to avoid shortages of energy and save money at the same time is to:
a. use less energy for everyday activities.
b. use secondary recovery methods to get more oil from old wells.
c. build more small-scale hydroelectric plants.

d. None of these choices will really succeed.

3. Ultimately, all the energy used on earth is derived from:
a. sunlight.
b. abiogenic production of organic compounds.
c. electromagnetism produced by Earth's rotation.
d. nuclear reactions in Earth's core.

4. The United States has 5 % of the world's population and uses ________ % of the world's energy, mostly from non-renewable energy sources.
a. 25
b. 75
c. 10
d. 50

5. When an industrial plant, such as a paper mill, produces steam for heat and electricity on-site, this is termed:
a. dual-fuel operation.
b. cogeneration.
c. simultaneous fuel utilization.
d. a power cooperative.

6. The measurement of energy used per unit time is called ____________________.
a. kilocalories
b. kilowatt-hours
c. power
d. first-law efficiency
e. second-law efficiency

7. One ____________ is equal to the heat needed to raise 1 g of water 1 °C.
a. joule
b. calorie
c. kilocalorie
d. BTU
e. kilowatt-hour

8. The total energy consumption in the U.S. in 1999 was ____________________.
a. 31 exajoules
b. 43 exajoules
c. 22 exajoules
d. 14 exajoules
e. 95 exajoules

9. Any process that uses heat to do work is called ______________________.
a. thermodynamically inefficient
b. a heat engine
c. second-law inefficient
d. a frictional force
e. high quality energy process

10. Which of the following systems has the highest level of entropy?
a. Gasoline in a car's fuel tank
b. Glucose in a candy bar
c. An incandescent light bulb that is illuminating a room
d. A stationary car at the top of a hill

11. The First Law of Thermodynamics states that:
a. Entropy within a system must increase when energy is converted from one type to another.
b. Energy may not be created or destroyed but may be converted from one form to another.
c. Matter can neither be created nor destroyed.
d. Energy equals mass times the square of the speed of light.

12. Which activity would best represent the "hard path" energy policy?
a. Carpooling to save gasoline and reduce automobile usage.
b. Building automobiles with increased gas usage efficiencies.
c. Increasing petroleum production to satisfy increasing demand.
d. Encouraging conservation by electricity consumers through economic incentives.

13. The force that converts kinetic energy into heat energy when work is done is called:
a. entropy.
b. power.
c. friction.
d. cogeneration.

14. Which of the following is an example of kinetic energy?
a. A bowling ball that is rolling down the lane toward some bowling pins.
b. A stationary bowling ball sitting at the top of the ball return rack.
c. The heating of the bowling ball by sunlight, as it sits in a window.
d. The energy stored in wooden bowling pins that could be used for heat by burning the pins.

15. Calculate the energy efficiency, according to second-law efficiency, if it takes 20 units of energy to actually do work that requires a minimum of 15 units to accomplish.
a. 133 %
b. 15 %
c. 20 %
d. 33 %
e. 75 %

Chapter 17 - FOSSIL FUELS AND THE ENVIRONMENT

The Big Picture

Fossil fuels (coal, oil, and natural gas) are highly concentrated forms of partially decomposed organisms that have been trapped in the Earth's lithosphere. They represent the storage of carbon compounds from the primary production and food webs that existed during the last 300 million years. These fuels now compose 90 % of our energy sources worldwide. We have rapidly used these resources in the past 100 years and we will have used them all, if we keep consuming them at current rates, in just 500 years. This may seem like a long period of time to an individual human, but this is an instant in geological time. Thus, in a relatively short period of time, we have released a great deal of carbon (CO_2) into the atmosphere that had been accumulating in the rocks for a long period of time. This rapid release of CO_2 has very serious implications for the human race and the climate (See Chapter 22 - The Atmosphere, Climate, and Global Warming). The other reality is that these fossil fuels are essentially non-renewable. The rate at which they are being produced naturally is slow relative to the rate at which we use them. In the past, we have been fortunate to find undiscovered reserves at a rate equal to our rate of use, but we don't know how much longer that will occur. We are bound to exhaust economically exploitable reserves sooner or later; the amounts of fossil fuels are finite. Eventually we need to discover alternative sources of energy. In this chapter, the authors discuss how fossil fuels were formed, how they are extracted, how large the reserves are for each fuel, and what environmental impacts occur during the extraction, transportation and use of fossil fuels.

Frequently Asked Questions

What are fossil fuels?

- The three main types of fossil fuels are coal, oil (or petroleum) and natural gas.
- Oil shale and tar sands are also fossil fuels.
- They are "fossil" fuels because they were formed a very long time ago (300 million years) by ancient organisms that died and have partially decomposed. They are fossils.

What is oil shale?

- Oil shale is a fine-grained sedimentary rock containing organic matter (kerogen).
- If heated to 500 °C, it will produce hydrocarbons (60 L/ton or 14 gal/ton).
- The production of petroleum from oil shale is called **synfuel** production.
- This practice is not currently economically feasible.

What are tar sands?

- Tar sands are sedimentary rocks or sands impregnated with tar oil, asphalt, or bitumen.

- Tar sands can be mined and heated with water to remove the oil.
- This practice is not currently economically feasible.

What are methane hydrates?

- Compounds containing methane (a natural gas) can be found a great depths beneath the seafloor.
- Potentially, methane hydrates could provide twice as much energy as all the known fossil fuel deposits.
- Presently methane hydrates are not economically or technically feasible to extract.

How were fossil fuels formed?

- **Oil & natural gas** - These two fuels are often found together in rock formations, so they were most likely formed in a similar way. Marine organisms, mostly marine diatoms, were deposited in marine sediments. After they had been covered with 500 m or so of sediment and rock layers during the last 300 million years, intense temperature and pressure have produced deposits of oil and natural gas, trapped in porous layers of rock, but under non-porous layers (Figure 17.2).
- **Coal -** Terrestrial and wetland vegetation became buried in sedimentary rock about 300 million years ago during the Carboniferous period (Figure 17.8).

How large are the reserves for each fossil fuel?

- **Oil** - There are about 1000 billion barrels of proven reserves.
    - Most (65 % by volume) of this oil is in the Middle East nations (Figure 17.3).
    - In 1972, there were only 350 billion barrels, so we have discovered about 650 billion barrels in excess of what we have used then.
- **Natural Gas** - 140 trillion m^3 of proven reserves.
- **Coal** - 1000 billion metric tons of proven reserves, mostly distributed between the U.S., Russia, and China (Figure 17.9).

What are resources and proven reserves?

- Undiscovered fossil fuels, which we suspect exist, but don't know for sure where they are or how extensive is the deposit are energy **resources.**
- A **proven reserve** is a subset of the resource but is known to exist and it what quantity and is economically exploitable.

How long can we use proven reserves of each kind of fossil fuel at current rates of consumption?

- We assume that use will continue at current rates of consumption; the actual rate of use may be lower or higher, depending on how many people there are and what people choose to do.
- We may also find more of each fuel; this will change the time at which we run out of fuel (the error estimate associated with these estimates is probably large).
- "Running out" means that it will no longer be economically effective to extract any more of the fossil fuel resource. We will never really extract all the oil, for example, because the cost will exceed the benefit at some point, leaving behind a minute amount of oil.
- **Oil** production is expected to peak in 20-50 years.

- **Natural gas** may last 70-120 years (30 years in U.S. alone)
- **Coal** may last 250 years but as other fuels decline, increased reliance on coal would decrease the supply.

How do oil and natural gas extraction and use affect the environment?

- The extraction of oil and natural gas is accomplished using similar methods.
- Of all the fossil fuels, natural gas is the preferred one to minimize environmental damage, because it burns cleanly into CO_2 and produces less CO_2 per unit of heat released than the others.
- The following is a list of environmental damages and changes that can occur:
    - **Land disruption** - Oil and gas wells, storage tanks, and roads must all be constructed; this disturbs native ecological communities.
    - **Surface and ground water pollution** - Broken pipes and ruptured storage tanks can contaminate water supplies. Water, steam or other chemicals are used to force out remaining oil after the well has stopped producing on its own; these can contaminate ground water. (Note: the oil and gas are found > 500 m deep, so that they are normally held in a different layer than ground waters used for drinking. Thus, contamination can happen when oil is drilled and the non-porous layer trapping the oil below is breached.) During refining process, or "fractional distillation" as it is called, groundwater contamination has been documented to occur (See Chapter 21 in text).
    - **Air pollution** - CO_2 is released from all fossil fuels when they are burned. Oil and natural gas both produce less CO_2 per unit energy than coal. Hydrocarbons are released into the air during extraction and refining of oil, and burning it as a fuel in cars and trucks. These airborne hydrocarbon compounds form urban smog (See Chapter 23 in text).
    - **Land subsidence** - As oil and gas are withdrawn, the land above the reserve sinks or subsides, because there is no longer anything occupying that space in the ground. This can cause property damage.
    - **Wildlife habitat may be disrupted**. - For example, The Arctic National Wildlife Refuge is near oil reserves in Alaska; for years there has been controversy over whether or not to allow drilling near this sensitive area. Recent congressional legislation has ruled against oil exploration but the issues is certain to arise in the future. In marine systems, oil drilling can have the impact of creating more habitat structure in the open ocean, which can actually attract fish to the rigs. The fish populations may be affected by having higher fishing harvest rates than prior to the rigs.
    - **Drilling mud releases** - Drilling "muds" or drilling fluids are used to lubricate the drill bit; they contain various mixtures of toxic (heavy metals) and non-toxic (barium) chemicals. These are often dumped over the side of the rig, contaminating the benthic environment around the rig.
    - **Aesthetic values are impaired** - Oil rigs are considered unsightly by some people.
    - **Oil spills during transportation** - Oil tankers spill oil into marine waters routinely. The pollution effects are short-term (10's of years), but the immediate impacts on

seabirds, marine fishes, invertebrates, and wildlife are devastating. Human communities suffer from lost tourism revenues and soiled beaches.

What are the energy contents and sulfur contents of the four types of coal?

- There are four types of coal:
  - **Anthracite** - High energy (30-34 J/kg), low sulfur
  - **Bituminous** - Intermediate energy (23-34 J/kg) , high sulfur
  - **Sub-bituminous** - Intermediate energy (16-23 J/kg), low sulfur
  - **Lignite** - Low energy (13-16 J/kg), low sulfur
- Low sulfur coal creates less air pollution (sulfur dioxide from burning coal makes acid rain; see Chapter 23).

How do coal extraction and use affect the environment?

- Coal mining is very damaging to ecosystems where coal is located (Figures 17.11, 17.12, 17.13).
- There are several negative environmental impacts of using coal for fuel:
  - **Strip mining** - This is the most commonly used method to mine coal. It involves removing the rock overlying the coal deposit (this is called the overburden) and thus it severely disrupts the land. Reclamation of land after strip mining takes a long time and has been successfully avoided by many coal companies. Only half of the land damaged by strip mining has been reclaimed. Unclaimed areas are visible from satellite imagery of Eastern Pennsylvania.
  - **Mine fires** - Coal is also mined by digging shafts down into the earth to extract the coal. Abandoned underground mines often catch fire. These mine fires continue to burn out of control in some areas, threatening life and property of the people above. They produce hazardous fumes.
  - **Land subsidence** - Subsidence occurs when the ground above a mine tunnel collapses, causing a pit to form. This occurs in Pennsylvania and West Virginia regularly. A parking lot and a crane fell in such a pit in Scranton, PA recently.
  - **Acid drainage from mines** - When coal mines and spoil banks are created, surface and ground water drains from the mines and dumps, bringing sulfuric acid into nearby streams and rivers. The Lackawanna River and the Lehigh River in eastern Pennsylvania have been made barren of aquatic life by acid mine drainage, although they are now both recovering.
  - **Air pollution from burning coal for electricity** - Coal burning releases 70 % of the total sulfur dioxides, 30 % of the nitrogen oxides, and 35 % of the CO_2 produced in the U.S.. It is thus a significant contributor to acid rain and global warming (See Chapters 22 and 23). Coal creates the most pollution of the fossil fuels.

How much CO_2 is released by gasoline fuel combustion?

- Gasoline for trucks and cars produces 25 % of the CO_2 released each year.
- Each gallon of gasoline burned creates 19 lbs of CO_2 to the atmosphere.
- An average car over its lifetime (100,000 at 27.5 mpg) emits 35 metric tons of CO_2.

What is a pollution trading allowance?

- A free-market approach called allowance trading for air pollutants has recently been introduced by the Clear Air Act of 1990.
- The U.S. EPA grants tradable allowances for polluting to utility companies that burn coal. One allowance unit is good for up to 1 ton of sulfur dioxide emissions per year. Pollution allowances can then be sold on a commodity market.
- If a utility company pollutes less than its allowance, it may sell the unused portion of the pollution allowance to another utility that pollutes more than its allowance. In this way the second company avoids a government fine.
- If a company invests in pollution control technology or changes its process to create less pollution, it may now profit from the reduction in pollution. There is a financial incentive for decreasing pollution.
- The overall level of pollution should not increase under such a scheme. It may even decrease, especially if environmental groups buy the pollution allowances first and remove them from the market.

Should gasoline taxes be raised?

- The suggestion has been made that if we increase the tax on fuels like gasoline by $1 per gallon, people will respond by using the fuels more conservatively (walk instead of drive, buy more fuel efficient cars, etc.). This tax would lower air pollution and decrease CO_2 emissions from automobiles.
- The revenues raised from such an increase would help pay for improvements in mass transit and could even pay off our national debt.
- Presently taxes in the U.S. on gasoline are low, about 30 cents/gallon; in European countries it is between $1 and $3 per gallon. Consequently, people in those countries drive much less than Americans.
- Some people are opposed to such an increase, because they believe that it would be unfair to people in western states who drive greater distances because of the sparse population. Westerners only spend 9 % more on fuel, however. These opponents also claim that poor people would be impacted, because they must drive and couldn't afford an increase. But poor people spend the same proportion of their income on transportation as rich people.
- This is a difficult social issue about which Americans are divided (in 1993 52 % favored a 15-cent increase, 46 % opposed it).

Ecology In Your Backyard

- Whenever you use electricity, most of that energy comes from a fossil-fuel electrical generation facility. Thus, by using electricity, we make CO2 and release it into the atmosphere. The Environmental Defense estimates that each person in the U.S. creates 40,000 pounds of CO_2 per year. Use the following chart to list all of the activities you do each day that cause fossil fuels to be burned and total up the savings in CO_2 that you make by being more conservative with energy:

Daily Activity or site for home improvement	Energy saving tip or appliance setting	CO_2 reduction pounds/year	Total for your home per year
Dishwasher	Wash full load, no heat energy-saving cycle	200	
Washer	cold water, not hot	500	
Water heater thermostat.	Set to 120 °F	500	
Water heater	Install insulating jacket	1000	
Thermostat	2 degree adjustment up in summer, down in winter	500	
Air filters in home	Replace when dirty	175	
Light bulbs	One compact fluorescent bulb in a frequently used place	500 per bulb	
Showering	Install low-flow shower heads	300	
Windows and Doors	Caulk and weatherstrip	1000	
Transportation	Walk, ride a bike, use mass transit instead of driving	20/gallon of gasoline	
Buying a car	Choose an efficient car (try to improve mpg by 10 over last car)	2500	
Solid waste	Reduce, reuse, recycle	1000	
Home insulation	Insulate walls and ceilings	2000	
Windows	Install energy saving ones	10,000	
Home exterior	Paint light in hot climate, dark in a cold climate, plant trees to shade it	5000	
Office/school papers	Recycle it!	4 /lb. of paper	
	Total CO_2 savings for the year in your home		

Links In The Library

- Corcoran, E. 1992. *Cleaning up coal.* Scientific American, May 1992. Can coal be made cleaner? The technologies are being developed now to do this, but this clean coal will be more costly and still cause damage from strip mining.
- Fulkerson, W. R.R. Judkins, and M. K. Sanghvi. 1990. *Energy from fossil fuels.* Scientific American, September 1990.
- Home Energy Magazine, 2124 Kittredge Street, No. 95, Berkeley, CA, 94704-9942. (510) 524-5405. Home Energy Magazine is a source of information on reducing energy consumption.
- Holloway, M. 1991. *Soiled shores.* Scientific American, October 1991. A report on the Exxon Valdez spill and the clean-up.
- Hubbard, H. M. 1991. *The real cost of energy.* Scientific American, April 1991. This article is about the subsidies provided for energy producers by the U.S. government and how market forces need to be allowed to act upon energy prices.
- Lenssen, Nicholas. 1993. *All the coal in China.* World Watch, March/April 1993, pp. 22-29. If China begins to use coal at the rate that other industrial nations have in the past, global warming will get a lot worse. China has abundant coal resources and intends to use them as it industrializes.

Ecotest

1. How old are most fossil fuels?
a. 30,000 years
b. 300,000 years
c. 3,000,000 years
d. 3,000,000,000 years

2. The type of fossil fuel that is the cleanest burning fuel is ___________________.
a. coal
b. oil
c. natural gas
d. tar sands

3. The size of the proven reserve for oil worldwide is ________________________.
a. 2000 barrels
b. 1000 billion barrels
c. 1 billion barrels
d. 350 billion barrels

4. Which fossil fuel can provide the world's energy needs for the longest time at current rates of consumption?
a. coal
b. natural gas
c. oil
d. shale oil

5. Which of these activities causes ground water pollution?
a. oil drilling
b. oil refining
c. oil recovery using steam or water
d. All of these are correct.

6. In terms of energy content, which of the following rankings of coal types is correct?
a. bituminous > sub-bituminous > anthracite > lignite
b. bituminous > sub-bituminous > lignite > anthracite
c. anthracite > bituminous > sub-bituminous > lignite
d. lignite > anthracite > bituminous > sub-bituminous

7. Which of these environmental effects of coal mining is so extensive that one can see it from remotely sensed satellite images?
a. mine fires
b. acid mine drainage
c. land subsidence
d. land disruption from strip mining

8. Each gallon of gasoline burned produces ____________ pounds of CO_2.
a. 100
b. 1000
c. 10
d. 19-20

9. Pollution trading allowances will:
a. prevent further increases in pollution emissions.
b. cause decreases in pollution emissions automatically through free-market incentives.
c. slow the rate of increase in pollution emissions.
d. prevent further increases and may lead to declines if allowances are bought and retired from the market.

10. Prices of gasoline are $3-4 per gallon in European countries, due largely to taxes. What have the taxes done to fuel consumption rates in those countries?
a. caused a reduction
b. caused an increase
c. no change
d. Fuel consumption rates have fluctuated due to factors unrelated to price.

11. Gasoline for trucks and cars produces _____ of the CO_2 released each year.
a. 50 %
b. 10 %
c. 25 %
d. 33%
e. None of these are correct.

12. Currently, it is not economically feasible to mine which of the following fossil fuels?
a. natural gas
b. coal
c. methane hydrates

13.Which type of coal would cause the most air pollution, because of its sulfur content?
a. bituminous
b. sub-bituminous
c. lignite
d. anthracite

14. The production of petroleum from oil shale is called:
a. cogeneration.
b. biofuel production.
c. oil simulation.
d. synfuel production.

15. The type of fossil fuel that produces the most air pollution is:
a. coal.
b. oil.
c. natural gas.
d. tar sands.

Chapter 18 - ALTERNATIVE ENERGY AND THE ENVIRONMENT

The Big Picture

Alternative energy sources to fossil fuels are available to meet the world's energy needs. These renewable sources of energy include solar power, geothermal power, wind power, water power, hydrogen fuel, and biomass fuels. These sources are all potentially large in available energy: each day, more solar energy hits the Earth than the entire world's human population could use in 27 years! It is not possible to use all of this available energy, however, because much of the sunlight falls on the ocean, where people do not live. But, even if a tiny fraction could be used to power our electric devices and machines, we could greatly reduce our reliance on polluting fossil fuels. Alternative energy sources will likely play an increasing role in US national energy policy as fossil fuel supplies decline and less pollution energy sources are sought. The major obstacle to instituting these alternative and renewable sources of energy is economic, not technological. Some communities are already increasing their reliance on alternative energy. For example, Pamplona, Spain obtains 23% of its electricity from wind-powered generators.

Frequently Asked Questions

What are the renewable energy sources?

- The renewable energy sources are solar (active, passive, direct, indirect, solar-thermal and photovoltaics), geothermal, wind power, hydropower, ocean thermal, hydrogen fuel, and biomass fuels.
- Making electricity from renewable sources normally involved heating water beyond the boiling point, where it can be used to spin turbines and generators. That is, it is the same as fossil fuel electricity production, except the way the water is heated is different.
- Most renewables (except biomass fuels) do not release CO_2 into the atmosphere, making them attractive from the standpoint of avoiding global warming. Even biomass fuels do not produce excess CO_2, because the plants grown to make the biomass fuel are taking CO_2 out of the atmosphere when they grow.

Why are renewable energy sources not used as much as fossil fuels?

- Economics are normally against them. Any new technology is costly at first, but the cost per unit of output drops over time.
- Eventually, these renewable sources of energy will be cost-competitive, especially if fossil fuel prices rise in response to scarcity and as a function of their pollution costs.

What are direct and indirect solar energy?

- **Indirect solar** is basically all forms of renewable energy - even fossil fuels are indirectly due to solar energy. Trees and plants are natural solar collectors. See Figure 18.2 for a breakdown of how all renewable energy can be traced back to the sun.
- **Direct solar** is using the sun's energy to heat water, air or make electricity at or near the point of use. There are several kinds of direct solar energy: Passive solar design, active solar water heating, active solar electric (photovoltaics and solar-thermal).

What are passive solar and active solar energy?

- **Passive solar energy systems** are normally designed into buildings when they are built. Simple design features include having windows face south with overhangs above the windows to block summer sunlight, but allow winter sunlight in.
- **Active solar systems** use pumps or fans to circulate the warmed water, fluids, and air in a building. An area is set up to collect solar energy and the heat is pumped away from that area to heat water for bathing, boil water to make electricity, or simply heat the air in a building.

How does passive solar design work?

- A building can be designed so that it catches the sun's light most of the day. In the northern hemisphere, that means having the major glazed surfaces facing south (Figure 18.4).
- Overhangs can prevent sunlight from entering during the summer when the sun is high overhead; in the winter, the sun is low in the sky, so sunlight can pass through the windows.
- The walls and floor in a south-facing room with many windows (called a sunspace) can be lined with brick, stone, tile or other masonry. The masonry walls will store the heat during the day and re-radiate it during the night. Water tanks will also store solar energy.
- Landscaping with trees can be used to minimize sunlight input during the summer and allow it to enter in winter (deciduous trees lose their leaves during the winter).
- Solar-heat collecting walls (glass enclosed brick or water filled **trombe walls**) collect heat and pass it through without a window being present.

What are the environmental costs and benefits of passive solar energy?

- **Benefits**: Low cost over life of building, although up front costs are high; little or no fuel required for heat, heating and summer cooling costs are lower than without the design; fossil fuel and CO_2 releases are minimized.
- **Costs:** The sun is an intermittent source of power (none at night or during cloudy days), so back up heaters are still needed. Thus total costs of the system exceed standard heating systems by 20 %.

How does active solar water heating work?

- Solar collectors are flat panels with glass covered cases with black interior surfaces (See Figure 18.5 in text).

- Water or anti-freeze (ethylene glycol) is pumped through a convoluted tube inside the enclosed airspace within the collector, and heat is absorbed by the fluid, reaching 38 - 93 °C (100-200 °F).
- The heated water can be used directly for showering or washing, or if anti-freeze fluid is used in the tube, the heat from the fluid is released within a large water heater tank and this water is used for household hot water.

What are the environmental costs and benefits of passive solar energy?

- **Benefits**: Although up-front costs are higher than a regular water heater ($1000 - $5000 versus under $1000), the system pays for itself in 7 - 10 years; little or no fuel is required to reach the 140 °F required by most dishwashers. Space heating and water heating costs (which are 25 % of utility bills) are much lower than without the design; fossil fuel and CO_2 releases are minimized (50 tons less of CO_2 over the lifetime of the equipment) .
- **Costs:** The sun is an intermittent source of power (none at night or during cloudy days), so back up heaters are still needed. Over the entire lifetime, a natural gas heater is still more economical (but it releases CO_2).
- In climates with mild winters and hot summers, buildings should be designed to dissipate heat as much during the summer and retain heat during winter. Air conditioning costs and energy use in many regions exceeds that of heating. Appropriate roof overhang, window glazing, ventilation, and landscaping are key to effective cooling.

What are photovoltaics (PV)?

- Photovoltaics are electricity-producing solar cells most people have seen on watches, calculators, satellites, remote lighting systems, and navigation buoys.
- Photovoltaics operate on the principle of excited electrons in solid semiconductor materials, which move from one material (silicon) to another (boron) in the presence of sufficient sunlight. Other cells are made of gallium arsenide.
- Panels are becoming standardized and mass produced.
- The amount of electricity produced is not great, but if the surface area covered with photovoltaics is large enough, they can produce up to 5000 kW of power.
- Utility companies are exploring the possibility of getting electricity this way "onto the grid." Most applications now are "off-the-grid."
- A new record of efficiency was recently set at 16.8 % of the sun's energy converted to electricity.

What are the environmental costs and benefits of photovoltaics?

- **Benefits:** There are numerous benefits; they could work forever, there is no maintenance once installed, no moving parts to break, free fuel (the sun) and minimal pollution during production and no CO_2 released during use.
- **Costs:** It's still more costly than fossil fuels or other forms of electricity to produce. During cloudy days and at night, 12 volt car batteries must be used to power lights and other appliances. All new electronics must be purchased that run on DC not AC (many can be converted). The batteries are recharged during sunny days, but the batteries need to be replaced and are filled with lead/acid, which causes lead pollution to increase. Some pollution will result during manufacture of the cells.

What is solar thermal electric power?

- Solar thermal electric is using the sun's concentrated energy to heat fluids to very high levels, and using it to boil water and make steam. Then a steam turbine generator is used to make electricity. There are three basic designs:
- **Parabolic mirrors** - The Luz Solar-electric generating system in the Mojave desert is the most successful attempt to date at generating large-scale electricity for the grid (Figure 18.10). It was successful because it combined natural gas heaters as a backup for nighttime and cloudy days. This insures uninterrupted power generation.
- **Power tower -** By having a tower with sun-tracking mirrors (heliostats) surrounding it, electrical energy can be generated in a manner similar to the parabolic mirror solar electric system. They have not proven to be as economically successful as the Luz system, but they do work to make electricity during the day.
- **Solar pond** -This is a shallow pond with a vertical salt water gradient, so that the denser, saltier water stays at the bottom of the pond and does not mix with the upper layer of fresher water. Consequently, the lower salty layer gets very hot (70 - 85 ^{o}C or 158 - 185 ^{o}F). This heat can be used to make electricity (with additional heating from traditional sources), provide energy for desalination, and to supply energy space heating in buildings.

What are the advantages and disadvantages of solar thermal?

- **Advantages:** Free fuel, no CO_2 releases
- **Disadvantages**: Needs large land area; not cost effective yet. But Luz was close to breaking the fossil fuel low-cost barrier.

Where in the U.S. is solar energy potential the greatest?

- The southwestern U.S. has the best solar potential, with over 26,000 kJ/m^2 solar energy hitting the Earth each year (Figure 18.3).

What is ocean thermal conversion?

- Ocean thermal relies on a temperature difference between surface waters (28 ^{o}C , 82 ^{o}F) and bottom waters (1 -3 ^{o}C, 35-38 ^{o}F) to generate heat or cause the expansion of ammonia gas that can turn a generator and make electricity. This energy source is largely in the experimental stage, so cost estimates are very high.

What is hydrogen fuel?

- In a hydrogen fuel cell, hydrogen gas is allowed to oxidize, forming water and releasing energy. So much energy is released that hydrogen gas can be explosive. In most applications, this is avoided by using an electrolyte solution in which to mix the gases; the result is an electric current that can run an electric motor.

What are the advantages and disadvantages of hydrogen fuel?

- **Advantages:** It is non-polluting; water is the only waste product; it can be used like a battery to store electricity made from any source, including photovoltaics and other solar sources. The electricity is run through water, producing hydrogen gas which is stored for later use. Can be used to power electric vehicles (Figure 18.11).

- **Disadvantages:** It is dangerous to have hydrogen gas around, so safe storage is a major hurdle.
- **Cost** will be $3 - $5 per gallon, similar to fossil fuels (taxes on gasoline need to be raised).

What is hydropower?

- Dams are created to store a head of water (potential energy), which is released through water turbines, which turn generators to make electricity. (Figure 18.12). A **micro-hydropower** facility is a small scale hydroelectric facility that can make a significant contribution to the electric needs of a small building or be sold "on the grid" to an electric cooperative.

What is tidal power?

- Similar to hydroelectric power, a head of water is created after the tide has entered an estuary or bay. The dam flood gates are then closed, and as the tide recedes, a water elevation difference occurs, head is generated, and water is released over turbines to make electricity.

What are the advantages and disadvantages of hydropower?

- **Advantages**: It does not pollute or produce CO_2. It is cost-competitive with fossil fuels.
- **Disadvantages:** The dams and reservoirs cause the deaths of many young anadromous fishes, especially salmon migrating downstream after hatching. These young smolts are susceptible to predators and nitrogen gas bubble disease. Also humans and wildlife are displaced when the river is flooded. Most of the places where a dam could be built in the U.S. already have hydroelectric dams. Dams also trap sediments, starving beaches of a sediment source and filling up the reservoir basin in the process. Tidal power can only work in a few estuaries where there is a sufficiently large change in tidal height.

What is wind power?

- Using wind energy to turn large propeller-like windmills and turbines. (Figure 18.14).
- There is great potential for wind energy and it has been installed in many places (The U.S. and Germany together account for more than half of the global wind energy production. Interestingly, Denmark, India, and The Netherlands rely heavily on wind power.
- China has great potential for wind energy and could possibly offset severe pollution problems associated with coal-fired power plants by using their wind resources.
- In the U.S., 80% of the wind power capacity is in just three windfarms in California. (Figure 18.15 in text).

What site conditions are necessary for wind powered generators?

- Site need a constant and steady wind velocity of 5m/sec.
- The most suitable sites include the Columbia River Gorge, Great Plains, Appalachian Mountains, southern California, and many coastal areas.

What are the advantages and disadvantages of wind power?

- **Advantages:** Cost competitive now with fossil fuels; no CO_2 or pollution during operation.
- **Disadvantages:** It can only be successful in certain constantly windy areas (although there are lots of such areas). Windmills kill migrating birds, including some endangered species like falcons, hawks, and bald eagles; vibrations and noise are produced; wind turbines can

interfere with radio and TV reception. They may also cause aesthetic values for an area to decline.

What is biomass fuel?

- Biomass fuels are those fuels that can be grown using the sun's energy to produce combustible products. Any organic material is potentially usable as fuel.
- For example, sugar cane can be grown and the sugar fermented to make ethanol. This "biofuel" can power automobiles. Other biofuels include methanol, methane or biocrude.
- Other biomass fuels include wood, wood pulp, peat, kelp, manure, and various kinds of agricultural wastes.
- Municipal solid waste can be burned for energy as well.

What are the advantages and disadvantages of biomass fuel?

- **Advantages**: Using waste products like solid waste, manure, and agricultural wastes is efficient. No net increase in CO_2 should occur, because plants remove CO_2 from the atmosphere when growing. Later, when they are burned, the CO_2 is replaced. If we used biomass instead of fossil fuels, CO_2 levels would decline.
- **Disadvantages**: Low net energy yield (it takes a lot of energy to grow crops and then ferment them for biofuels); produces CO_2, although no net increase of CO_2 should occur. Using wood biomass for fuel reduces the availability of forest products for more traditional uses.

What is geothermal electricity?

- Geothermal is the use of earth heat or heated rocks to heat buildings or make electricity.
- In **geothermal electrical generation**, water is pumped into the Earth, the temperature of water is raised above the boiling point, which creates steam that can be used to spin a turbine and a generator to make electricity.
- Geothermal can be considered a non-renewable resource if heat is extracted from the rocks faster than it is produced by the Earth. This is happening at the Geysers Geothermal facility in California (Figure 18.18), which is experiencing a decrease in geothermal energy output each year.
- Geothermal resources occur in specific areas of the world, where the Earth's tectonic plates collide.
- There are 2,700 megawatts of geothermal energy produced worldwide, about 2,000 MW comes from the Geysers plant alone. The Geysers produces about 7 % of California's energy needs.
- In **geothermal groundwater systems**, a geothermal heat pump is used to warm the air in a building. The temperature difference between the surface air and the ground water is usually great enough in the summer to cool a home and warm enough in the winter to warm it. Ground water stays a constant 13 ^{o}C (55 ^{o}F) all year, but the air temperature fluctuates.

What are the environmental benefits and costs of geothermal electricity?

- **Benefits:** Relatively clean, no CO_2 emissions, free heat source, no transportation of fuel
- **Costs:** Noise, gas emissions, may be non-renewable, hot wastewaters are thermal pollutants and are often corrosive. Many Hawaiians believe in Pele, the goddess of fire, and think this

practice offends her and "steals her breath and water". This opposition is based on culture, not science, however.

- **Cost per kilowatt-hour:** \$0.05 - \$0.075

Which renewable energy technology has the most promising future in meeting our nation's energy needs?

- That would be wind or geothermal, because they are competitive now. Solar thermal is promising as well, but it is currently more costly than fossil fuels.

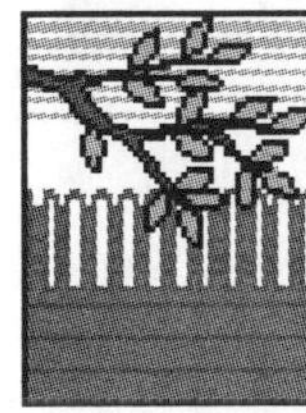

Ecology In Your Backyard

- Can you make use of more alternative energy in your home or lifestyle? For example, try to purchase solar-powered electric devices whenever you can (calculators, watches, lighting for the outside of your home, personal digital assistants, etc.). The price is usually a little higher than a similar battery-powered product, but you will save money on battery purchases in the long run and not create a toxic waste problem when you dispose of the batteries after they are exhausted.
- Look for the U.S. EPA "Energy Star" label to find the most energy efficient product when you buy anything that consumes electricity or fossil fuels.
- When you design or purchase a new home, do you consider the energy savings that can be obtained by using passive solar design? That is, will your new home save energy by having the glazing of the home facing south (in the northern hemisphere) to take advantage of the sun's (free) energy? You will need to have overhangs or awnings installed to cut down on the summer sun (high angle) as well, but during the winter your home will be warmed by the sun.
- Can you modify your existing home to include more solar power, including adding an after market solar hot water heater or photovoltaic equipment? These devices are typically more expensive in the short run than their fossil fuel equivalents, but in the long-term (more than 8 - 10 years) they are cost competitive or less expensive than fossil fuels. You won't produce nearly the same amount of pollution, however, and CO_2 emissions are dramatically reduced. This helps prevent global warming.
- Purchase a solar heated water bag (a Sunshower™ or similar product, available at most outdoor stores or in the camping section of many discount department stores) and use it the next time you go sailing, to the beach, or camping. The Sunshower is a simple design: it consists of a plastic bag with a clear side and a black side, a port for filling it, and a tube with a low-flow shower head. The water you put in the bag (about 5 gallons) will be very warm (110 ^{o}F) within a few hours. It's great for washing off sandy feet at the beach or providing hot water for washing while camping. I use mine all the time when camping.

Links In The Library

- Keisling, B. 1983. *The Homeowner's Handbook of Solar Water Heating Systems*, Rodale Press, 1983.
- Manning, R. 1994. A Good House: Building a Life on the Land. Penguin, New York, NY.
- Rosenbaum, M. 1991. Solar hot water for the '90s. Solar Today, September/October 1991, 5(5): p. 20.
- Sklar, S. and Sheinkopf, K. 1991. *Consumer Guide to Solar Energy*, Bonus Books, Inc., 160 East Illinois Street, Chicago, IL 60611, 1991.
- Solar Industry Journal, Solar Energy Industries Association, 122 C Street, NW, 4th Floor, Washington, DC 20001. Solar Industry Journal has information on commercializing new technologies, case studies of commercially available technologies, and articles on government policies and regulations that affect renewable-energy businesses.
- Solar Today, 2400 Central Avenue, Unit G-1, Boulder, CO, 80301. (303) 443-3130. Solar Today covers all the solar technologies, both mature and emerging, in a general-interest format.

Ecotest

1. Each day, more solar energy hits the surface of the Earth than the entire human population of Earth could use in ____________________.

a. 15 days
b. 27 weeks
c. 27 years
d. 10 weeks

2. Which of the following renewable fuels releases CO_2 into the atmosphere in the process of making usable energy?

a. Passive solar
b. Photovoltaics
c. Hydropower
d. Wind power
e. Biomass fuels

3. It has been said that all we are doing when making electricity is "boiling hot water." For which of these renewable energy sources is this not true?

a. Solar -thermal
b. Photovoltaics
c. Geothermal electric

d. Biomass fuels

4. Why was the Luz International Company's solar design different?
a. They used a tower power to generate electricity.
b. They used a natural gas heater to augment the solar energy on cloudy days and at night for a continuous output.
c. They used passive solar design throughout.
d. Photovoltacic pumps were used to pump the hot oil through the tubes.

5. The solar energy potential of the ____________________ United States is the best for large scale solar energy.
a. Northeastern
b. Northwestern
c. Southeastern
d. Southwestern

6. A temperature difference between surface and bottom water can be used to pump ammonia down and allow it to expand rapidly as a gas. This expanding gas drives a turbine and makes electricity in a process known as ________________________.
a. solar pond thermal electric
b. hydrogen fuel cells
c. tidal power
d. ocean thermal conversion

7. Which of these is not a disadvantage of hydropower?
a. Dams cause anadromous fish deaths.
b. River basin is flooded, displacing wildlife.
c. The vibrations of thc turbines produce radio and TV interference.
d. Sedimentation is filling in behind the dams.

8. If both release CO_2, why is biomass fuel preferred by environmentalists over fossil fuel?
a. It reduces the build-up of organic waste products.
b. There is no net increase in CO_2 in biomass.
c. Biomass fuels can be grown in areas unsuitable for human food production.
d. Biofuels are cleaner burning than all fossil fuels except natural gas.
e. All of these are correct.

9. What is a major disadvantage of wind power?
a. It kills endangered birds.
b. The windmills produce excess CO_2.
c. The equipment is unreliable.
d. Not cost competitive with fossil fuels.

10. ________________ is the type of design in which the building's design is planned in advance to optimize free solar energy to be used as heat.
a. Indirect solar

b. Direct solar
c. Active solar
d. Passive solar

11. The record for conversion efficiency of solar energy into electricity using photovoltaics stands at:
a. 2.5%
b. 16.8 %
c. 52.1 %
d. 98.6 %

12. A major disadvantage of hydrogen fuel is:
a. it is much more polluting than fossil fuels
b. it does not have much ability for storage of energy
c. it has limited practical uses
d. it is dangerously explosive

13. All of the following are examples of biomass fuels except:
a. ethanol created from fermented sugar.
b. municipal solid waste.
c. wood.
d. tar sands.

14. Which of the following was closest to the average price of electricity per kilowatt-hour at the Luz International solar electric company?
a. $0.01
b. $0.10
c. $1.00
d. $5.00

15. Groundwater, in temperate climates, maintains a constant annual temperature of approximately:
a. 13°C
b. 32 °C
c. 55 °C
d. 10°C
e. 0 °C

Chapter 19 - NUCLEAR ENERGY AND THE ENVIRONMENT

The Big Picture

Perhaps no other area of energy policy has engendered more debate and controversy than nuclear power. The mere thought of a nuke plant down the road spewing out dangerous radiation strikes fear in many people's minds due to Chernobyl, Three Mile Island, The China Syndrome, and our memories growing up of the horrible scenes from Hiroshima and Nagasaki. In reality, the nuclear energy technology is relatively safe compared to other technologies that we use every day, like automobiles. In fact, most of the radiation we are exposed to every does not come from a nuclear power plant, but from nuclear medicine and x-rays done for medical purposes. For instance, about the same amount of radioactivity comes from the smokestacks of coal-burning electrical power plants as comes from nuclear plants; both of these are lower than the dose the average person gets from doctor's x-rays in a year. Nuclear plants have not caused the death of anyone in this country of which we can be certain. Some of the reasons people fear nuclear energy technology are that they do not understand the units of measurements for radiation, what a half-life is, what the different kinds of radiation are and what each of their biological effects are, how nuclear energy is produced, what the nuclear fuel cycle is, how much radiation is released normally from a nuclear plant, and what we have learned from accidents at nuclear plants. In this chapter, the authors discuss each of these issues and the science underlying them. They also discuss the present current thinking on how nuclear waste may be safely disposed of, and what the future of nuclear power is likely to be.

Frequently Asked Questions

What is nuclear fission?

- A neutron is used to split an atom's nucleus (usually uranium 235) into smaller nuclei (radioactive isotopes that are the breakdown products of U-235), which causes the release of energy and three neutrons.
- These neutrons break away from the nucleus and then each strikes another U-235 nucleus. Now these three U-235s are split, releasing 3 more neutrons each. These nine neutrons strike more U-235 nuclei and release 27 neutrons. This nuclear chain reaction, if unchecked, can proceed rapidly into a nuclear explosion (Figure 19.1).

What is fusion?

- Fusion is the joining together of two molecules of hydrogen (H) to form helium (He). This fusion reaction would combine two atomic nuclei with one proton and one neutron each to form another atomic nuclei with two protons and two neutrons. This releases lots of energy. Helium is not a pollutant (Figure 19.6).

- This is the reaction that occurs on the sun. Although it has been a goal of some very high-tech research programs for some time, fusion has only been attained for a few seconds. These scientists contained the reaction in a magnetic field at extremely high temperatures.
- If this form of nuclear power could be achieved, it would be non-polluting and endless. But it has not been achieved yet on a commercial scale.

What is a radioactive isotope?

- A chemical element that spontaneously undergoes radioactive decay.

What are the isotopes of uranium?

- There are three isotopes in uranium ore. The normal percentages of the isotopes in the ore are 99.3 % U-238, 0.7 % U-235, and 0.005% U-234.
- U-235 is the only fissionable isotope but at 0.7% the concentration is too low for nuclear chain reactions to occur.
- With special purification and centrifugation procedures, the U-235 can be concentrated to 4 % of the sample. At this concentration of U-235, the fission reaction starts to accelerate rapidly on its own in a positive feedback loop.
    - There is a limited supply of U-235. If the 440 nuclear power plants in the world today increased the their total contribution of the worlds electricity from 16% to 40%, the supply of U-235 would be exhausted in 30 years.

What is radiation?

- Radiation is the process that occurs when a radioactive isotope decays into another isotope and releases particles.
- There are three kinds of radiation particles:
    - **alpha particles** - Consist of two protons and two neutrons (a helium nucleus); this particle has the greatest mass, but it will not travel far (5-8 cm in air); little cell damage occurs due to this type of radiation. Toxic if ingested.
    - **beta particles** - These are electrons, and they have little mass (1/1840 the mass of a proton). They travel farther than alpha particles, but they can be stopped by wood or metal.
    - **gamma rays** - This is the most harmful type of radiation. It consists of electromagnetic radiation, similar to an x-ray. This can penetrate thick shielding and human bodies, but is stopped by lead.
- None of these are visible or can be sensed in any way by humans, unless Geiger counters are used to monitor them.

What are the units of radiation?

- Curies (Ci) - unit of radioactive decay = 37 billion nuclear transformations/second
- Becquerel (Bq) - unit of radioactive decay = 1 radioactive decay per second = 1 trillionth of a curie
- rad (rd) = radiation absorbed dose
- 1 Gray = 100 rads (these are both units of the absorbed dose of radiation)
- rem = effective equivalent dose; = (relative biological effectiveness of radiation x dose in rads)

- 1 sievert = 100 rems (these are both units of effective radiation dose, after accounting for the biological effectiveness or damage that is caused by different types of radiation)
- roentgen = unit of gamma radiation in coulombs/kilogram (a coulomb is a unit of electrical charge)

What is the naturally occurring or background radiation received by people?
- 1.5 millisieverts/year (1.0 - 2.5 millisieverts/year) total. (1 sievert = 1000 millisieverts)
- The average person in the U.S. will receive the following doses from each source:
  - Medical X-rays: 0.7 - 0.8 millisieverts/year
  - nuclear power plants/weapons tests: 0.04 millisieverts/year
  - coal, oil, natural gas burning: 0.03 millisieverts/year
- See radiation dose charts in text (Figure 19.10 and 19.11)

What are the environmental and human health effects of radiation?
- A dose of 5000 millisieverts(5 sieverts) is lethal to 50 % of people.
- A dose of 1000-2000 millisieverts (1 -2 sieverts) is damaging to human health (temporary sterility in males, aborted pregnancy, vomiting, fatigue, and other problems).
- The maximum permissible dose set by the U.S. NRC for people is 5 millisieverts per year.
- Workers that mine uranium have an elevated rate of lung cancer.
- It is uncertain if low levels of radioactivity can cause cancer, because it depends whether a linear or non-linear dose response curve for radiation is used. Because it is not known for certain, it is best to assume the worst case and use a linear model. This means that no matter how low the dose is, some cells may become cancerous.
- Radioactive materials can biomagnify in the food chain (Figure 19.13).

How is electricity made from nuclear fission?
- The rapid fission reaction can lead to a rapid release of energy that is used at the core of the nuclear energy generation procedure: this release of heat and energy is what is used to heat water in the nuclear reactor.
- The heated water is used to create superheated steam, which spins a steam turbine, and this turns a generator that makes electricity.
- The intensity of the nuclear fission reaction is controlled by lowering control rods in a nuclear reactor, which stop the neutrons from continuing the chain reaction.
- This is similar in principle to a coal burning plant, except that the water is boiled using a nuclear fission reaction, not burning a fossil fuel (Figure 19.4).
- Approximately 1/5 of the electricity produced in the US is from nuclear power plants.

What is a nuclear reactor?
- This is the location in the nuclear facility where the nuclear fission reaction is contained (Figure 19.5).
- The main components of a reactor are the core (fuel and moderator), control rods, coolant, and reactor vessel.
- The core of the reactor is enclosed in a stainless steel reactor vessel; the reactor vessel is contained within a reinforced concrete building.

What is a burner reactor?

- This is a nuclear reactor that consumes more fuel than it produces.
- This is the most common type of reactor in nuclear plants.
- The nuclear reaction is a fission reaction.

What is a breeder reactor?

- This is a reactor that produces more fissionable fuel than it consumes.
- This type of reactor can use waste or low-grade uranium more efficiently than a burner reactor, thus extending the world's uranium supply.
- Breeder reactors are not technically feasible today and are not in use.

What is a meltdown?

- This is a nuclear accident in which the nuclear fuel becomes so hot that it melts the stainless steel reactor vessel, contaminating the outside environment.
- Meltdowns have not occurred in the past, but it is a possibility if the coolant leaks out or if the coolant pumps should fail.
- New reactor designs can prevent meltdowns by using small nuclear fuel grains instead of rods . These can't get hot enough, even with no coolant at all, to cause a meltdown.

What nuclear power generating accidents have occurred?

- Two major accidents have occurred, with many more minor incidents.
- **Three-Mile Island accident, Harrisburg, PA, 28 Mar 1979**
    - Malfunctioning pumps caused a partial meltdown in the reactor core.
    - Releases of radioactive isotopes were allowed into the environment.
    - 1 millisievert was released (a low amount of radioactivity), and average exposure was 0.012 millisieverts in the entire area around the plant; some readings near the plant were as high as 12 millisieverts/hour. Thus, exposure to radiation was variable.
    - There is a possibility that some people were exposed to higher levels than others and that some excess deaths from cancer may have occurred, be we cannot separate the victims from cancer cases caused by other things.
- **Chernobyl, near Kiev, USSR, 26 Apr 1986**
    - This was the most serious nuclear accident ever. At first it was not announced by the Russians, only to be discovered by Swedish nuclear monitors 2 days later and some distance away.
    - It has been suggested that the cooling system of this plant failed, which caused a complete meltdown of the graphite-core reactor. A fire broke out in the reactor core, blowing the roof off the building.
    - This facility had a different design than the nuclear plants in the U.S.. It used a graphite core, which can lead to the problems descried above if a coolant system fails.
    - More than 3 billion people in Europe and North America were exposed to some level of radiation from this event; the most significant exposures for human health occurred within 20 miles of the plant. This area has been evacuated by humans.
    - 31 people died and 237 people suffered from radiation sickness.
    - Leukemia and other cancers may increase in the future among the people exposed to high levels within 20 miles of the plant. Already, 653 thyroid cancer cases have been

described in children around the plant. It has been projected that in the future there may be as many as 16,000 deaths from cancer attributed to this event.
- Vegetation was killed or severely damaged within a 7-km radius of the plant.
- The populations of animals (voles, otters, moose, waterfowl and wild boars) have increased since the accident in the evacuation zone. This increase is puzzling, because the rodents tested have shown an increase in mutation rates, yet their populations are greater than before. The increase is likely to be due to the lack of humans in the area, and the radiation is not great enough a dose to kill them immediately.

What is the nuclear fuel cycle?
- This is how the nuclear fuels are mined, processed to become fissionable, reprocessed after use in a reactor, and finally stored or disposed of (Figure 19.8).
- Radioactive releases can occur at any point in this cycle, but are especially common at the mining, disposal, and decommissioning stages.
- Decommissioning nuclear plants after 30 years of operation (the useful lifetime) will be an expensive part of the cycle, and one that we have little experience with.

How is nuclear waste currently handled?
- There are three kinds of radioactive waste: low-level, transuranic and high-level waste.
- **Low-level waste** is currently disposed of by burial at two remaining low level sites (in WA and SC). Most low-level waste is liquid waste from power plants, radioactive chemicals, and contaminated equipment.
- **Transuranic waste** is also low-level radioactive and is composed of man-made radioactive materials. The waste usually consists of contaminated industrial trash, clothing, tools and equipment.
- **High level waste** is temporarily stored in pools of water on the site where it is produced.
- There are about 29,000 tons of radioactive waste and spent fuel rods currently in temporary storage.

What will happen to high-level nuclear waste in the future?
- Permanent storage of high-level waste produced in the U.S. is being considered at an isolated underground storage site at Yucca Mountain , NV.
- The site is believed to be geologically stable but the waste deposited there will remain dangerous for 100,000 years or more.
- There is strong local opposition to opening a national depository at Yucca Mountain and there are concerns about the potential accidental exposure of great numbers of people to radiation as wastes are transported from the nations power plants to the facility in Nevada.

What are the disadvantages of nuclear energy?
- The decommissioning issue and radioactive waste disposal issue is unresolved.
- In order to achieve sizable reductions in the release of CO_2 from coal plants, nuclear power industry would have to expand tremendously their electric generation capacity and this goal would not be reached for more than 50 years. In that time, solar-electric could become economical.

- New nuclear plants are expensive because of safety concerns.
- Politically, nuclear power is very unpopular.
- In the U.S., no new nuclear power plants have been ordered since 1978.

What are the advantages of nuclear energy?

- It does not pollute the air or cause acid rain like fossil fuels do.
- It does not produce CO_2 or contribute to global warming.
- If breeder reactors are developed, they will extend the available fuel supply, making it inexhaustible.
- It is safer than other forms of energy for people; with the exception of the two major accidents outlined above, no one is thought to have died from this activity.

Ecology In Your Backyard

- Where is the closest nuclear power plant to your home or school?
- After reading this chapter, has your opinion about nuclear power changed?
- Would you allow a low-level radioactive waste dump to be built in your community?
- Would you allow high-level radioactive waste to be transported through your community?

Links In The Library

- Craxton, R. S. et al. 1986. *Progress in laser fusion*. Scientific American 255(2): 68.
- Ginsburg, S. 1996. *Nuclear Waste Disposal, Gambling on Yucca Mountain*. Aegean Park Press, Laguna Hills, CA.
- Goldsmith, E. et al. 1986. *Chernobyl: the end of nuclear power?* The Economist 16(4/5): 138-209.
- Hatele, W. 1990. *Energy from nuclear power*. Scientific American 263(3)
- Lester, R. K. 1986. *Rethinking nuclear power*. Scientific American. 254(3): 31.

Ecotest

1. How many uranium 235 atoms will be split by the neutrons arising from each collision in the nuclear chain reaction?
a. 1
b. 2
c. 3
d. 4
e. 5

2. What is the difference between nuclear fusion and nuclear fission?
a. In fusion, hydrogen atoms are split; in fission, helium atoms are fused.
b. In fusion, hydrogen atoms are fused; in fission, uranium-235 atoms are split.
c. In fusion, helium atoms are fused; in fission, uranium-235 atoms are split.
d. In fusion, uranium-235 atoms are fused; in fission, helium atoms are split.

3. Which type of radiation particle is most harmful to humans?
a. alpha
b. beta
c. gamma
d. delta

4. The radiation dose which will kill 50 % of humans exposed to it is 5000 millisieverts/year (we know this from the atomic blasts at Hiroshima and Nagasaki). What is the dose of radiation that the average person is exposed to annually from all sources (including nuclear power plants)?
a. 10,000 - 20,000 millisieverts/year
b. 1000 - 2000 millisieverts/year
c. 100 - 200 millisieverts/year
d. 10 - 20 millisieverts/year
e. 1 - 2 millisieverts/year

5. How do a coal-burning and a nuclear power plant differ?
a. The coal plant releases CO_2 and no radiation, whereas the nuclear plant releases radiation and CO_2.
b. The coal plant releases no radiation, but the nuclear plant does.
c. The coal plant boils water with a coal fire, the nuclear plant boils water with the heat from a nuclear fission reaction.
d. All of these are correct.

6. Which of these was these was associated with the Chernobyl nuclear accident?
a. A fire broke out in the reactor core, causing a steam explosion that blew the roof off the building, releasing radioactivity into the environment, and killing 31 people.
b. There was a partial meltdown of the reactor core due to malfunctioning pumps.
c. 1 millisievert of radiation was released.
d. All of these are correct.
e. None of these are correct.

7. In the U.S., high-level nuclear waste is currently stored in ______________________________.
a. Washington, DC
b. pools of water at each nuclear power generating facility
c. South Carolina
d. Washington State
e. Yucca Mountain, Nevada

8. Which of the following is a disadvantage associated with nuclear power?
a. We will run out of uranium in 10 years.
b. Nuclear waste generates the precursors to acid rain.
c. Excessive CO_2 is produced, which causes global warming.
d. The decommissioning of nuclear power plants and the safe disposal of nuclear waste are unresolved.
e. All of these are disadvantages.

9. Which of the following are advantages associated with nuclear power?
a. It does not pollute the air with particulates, sulfur and nitrogen compounds like fossil fuel plants.
b. It does not produce CO_2 like fossil-fuel burning plants do.
c. It is safer than other sources of energy production.
d. It is less costly at present than other forms of energy that do not generate CO_2.
e. All of these are correct.

10. A__________________ reactor produces more fuel than it uses.
a. nuclear fission
b. burner
c. nuclear fusion
d. breeder
e. None of these are correct.

11. If all of the nuclear reactors online today were to increase power production to 40% of the total electricity produced, the supply of uranium-235 fuel would be exhausted in:
a. 3 years
b. 30 years
c. 300 years
d. We will never run out of uranium.

12. All of the following are units used to measure radiation except:

a. Becquerel
b. Curies
c. roentgen
d. sievert
e. Geiger

13. The dominant isotope in uranium ore is:
a. U-235
b. U-236
c. U-237
d. U-238

14. The unit of radiation called the "rad" measures:
a. radioactive decay.
b. absorbed dose of radiation.
c. the biological damage that is caused by different types of radiation.
d. the amount of power created by the fission process.

15. The fusion of two hydrogen atoms results in:
a. a hydrogen molecule, H_2
b. a helium atom, He
c. an oxygen atom, O
d. a carbon atom, C

Chapter 20 - WATER SUPPLY, USE, AND MANAGEMENT

The Big Picture

Water has unique physical properties such as its high specific heat, its strong surface tension, its lower density as a solid than in the liquid form, and the fact that it is a universal solvent. Water is unevenly distributed temporarily and spatially. Water in its various states (ice, liquid water, and water vapor) is found in the atmosphere, the oceans, and freshwater lakes, rivers, and reservoirs, and in the ground. Transport among these compartments forms the hydrologic cycle, which is solar-driven. Water budgets are an accounting of water inputs to and outputs from systems. The attempt by humans to control the hydrologic cycle is called water management. Poor water management by humans may result in detrimental environmental consequences. Users of water have varied over the years and often have incompatible value systems. Key water-related issues discussed in this chapter include wetland alteration, dam and reservoir construction, stream channelization and aquifer overdraft.

Frequently Asked Questions

What are some of the unique or uncommon properties of water?

- Water is essential for life.
- Water has a high specific heat, thus does not change temperature rapidly. This enables organisms to successfully regulate temperature.
- Water is the "universal solvent." It is capable of dissolving a great variety of substances, from salt (rapidly) to granite rock (slowly).
- Nutrients dissolve and are mobilized by water.
- Water has a high surface tension, which facilitates movement through membranes and vascular tissue.
- Water in its solid form (ice) is less dense than liquid water, thus ice floats.
- Water is transparent, thus capable of transmitting light for photosynthesis .

How much water is there and how is water distributed globally?

- 97% of the Earth's water is in the oceans, of course, this is saline and of limited direct use by humans (Table 20.1).
- 2% of the Earth's water is in polar ice caps and glaciers. This is freshwater but is not readily accessible for human usage.
- The remaining 1% is located in groundwater, surface water (rivers, streams, lakes), and the atmosphere.
- Plants, animals, and microbes, and freshwater aquatic organisms depend on this one percentage of global water.
- The global distribution of water is not uniform (Table 20.2).

- Much of the global precipitation occurs far from human population centers and either evaporates or runs off and is not available for human usage. However, this "unused" water drives important ecosystems such as the Amazonian rainforest and the boreal forest of Canada.

How much water do humans use?
- Humans rely primarily on groundwater and surface water supplies for domestic, agricultural and industrial use.
- Global per capita usage in 1975 was 700 m^3/yr (185,000 gal/yr) or 2000 liters/day.
- Estimated per capita usage by year 2000 will be 6000 m^3/yr.
- Per capita water usage varies greatly among nations. The more developed nations which have high industrial and/or agricultural output consume far more water than the less developed nations.

What is a water budget?
- A water budget is an accounting of the water present in the various compartments of a system and the pathways between these compartments.
- The hydrologic cycle (Figure 20.2) is a depiction of water pathways.
- Water cycles through the compartments of the hydraulic cycle but the residence time in each compartment varies from days (in the atmosphere) to thousands of years (in the oceans or glaciers).
- Water budgets may be developed for local ecosystems (natural and human-dominated) as well as globally (the global hydraulic cycle) (Table 20.2)

What constitutes water quality and water quantity?
- Water quality is a measure of the physical, chemical, and biological conditions of a water body. Pollution causes water quality to deviate from normal conditions. Water quality is discussed in Chapter 21.
- Water quantity (or water supply) pertains to the delivery and volume of water.

What are the sources of our water supply?
- Most of the water used by humans comes from groundwater aquifers or surface water.
- Water infiltrates groundwater systems from recharge zones.
- Aquifers are underground soil or rock bodies that contain pore spaces or fractures which may become filled with water.
- Aquifers nearest the ground surface are called surficial or unconfined aquifers. The top of the surficial aquifer is the water table. These aquifers are especially vulnerable to pollution.
- If the ground surface is below the potentiometric water table of an aquifer, then water from the aquifer discharges to the surface. These discharge zones appear as springs, seeps, wetlands, ponds and lakes, and groundwater-fed streams.
- Deeper aquifers, which are separated from the surficial aquifer by an impermeable layer, are called confined aquifers.
- Streams that discharge water into aquifers are called influent streams; streams that receive discharge from aquifers are called effluent streams.

What is the fate of precipitation?

- An average 66% of total precipitation either evaporates or is transpired by plants, thus does not enter our water supply.
- Evaporation and transpiration rates vary depending upon season, vegetation, soils, land use, and local climatic conditions.
- The remaining 34% of total precipitation enters groundwater or surface water supplies.
- Water supply shortages are most likely to occur in areas of low precipitation and runoff, and water usage patterns that exceed sustainable water supply.
- Water usage in the United States is projected to exceed water supply by 13% by the year 2020.
- Droughts often occur in cycles and may last for years. Droughts are most acute where water use is incompatible with available water supply.

What is groundwater overdraft?

- Groundwater is a primary source of drinking water for about half of the U.S. population.
- The total groundwater supply of the U.S. is considerable; groundwater volume within 0.8 km of the surface is estimated to be 125,000 to 224,000 km^3.
- However, groundwater is unevenly distributed and withdrawals exceed recharge in some regions. This excessive removal of groundwater is called overdraft.
- The Ogallala aquifer of the High Plains region of Texas and Oklahoma has experienced significant overdraft due primarily to agricultural irrigation (i.e., center pivot irrigation).

Can seawater be used as a water source?

- Seawater can be desalinated but the process consumes much energy and is expensive.
- If no other water sources are available, desalinization may be a viable option.
- As with mining any resource, water or mineral, extraction depends upon technological and economic feasibility (see Chapter 29).
- Coastal nations with abundant economic and energy reserves and a scarcity of water, such as Kuwait, depend largely upon desalinated water.

What are the major use categories of water?

- Off-stream uses are those uses that temporarily or permanently remove water from its source, such as using water to irrigate farmland or as a coolant in power plants. If the water is not returned to its source, the use is categorized as consumptive use, such as removal for domestic consumption.
- In-stream uses are those uses that do not remove water from its source, such as hydroelectric power generation, navigation, recreation, and maintenance of fish stocks for commercial and sports fishing.
- The multiple uses of water are sometimes incompatible and conflicts among users result.

How is water transported from its source to its site of use?

- Population centers often lack a sufficient water supply and must obtain water from distant sources.
- Additionally, water supplies in the vicinity of population centers often become polluted, further necessitating reliance on distant water supplies.

- To retain adequate water supplies, dams and reservoirs are constructed.
- Water may be transferred from distant areas by means of canals and aqueducts.
- Cities in semi-arid, central and southern California must rely on distant water sources located in the Sierra Nevada Mountains. An extensive canal network (the California Water Project) delivers water to these cities.
- New York City, even though located in a humid region, must obtain water from distant reservoirs. Local supplies do not adequately meet demand and are too polluted.

What trends are occurring with water use in the United States?
- From 1950 to 1980 water use steadily increased (Figure 20.10). Human population has increased concurrently.
- From 1980 to present, population has continued to increase but water use has remained relatively constant.
- This trend suggests that water conservation and management efforts have been somewhat successful.
- However, obtaining an adequate water supply will be an on-going problem especially in high population density areas in semi-arid and arid localities.
- Water for irrigation and the thermoelectric industry are the largest uses of water (Figure 20.11).
- Water usage for domestic consumption (public supply and rural) has increased as population has increased, but industrial consumption has decreased.

As demand for water increases, how can we ensure an adequate supply in the future?
- Since 80% of the water used in the United States is for agricultural irrigation, water conservation in this sector should be a high priority. Conservation efforts should include:
  - appropriate pricing for water usage
  - reducing losses from canals
  - improved management for water delivery
  - combined use of surface water and groundwater supplies
  - irrigating more efficiently
  - appropriate crops
- Domestic use accounts for only 10% of the national total use, but in densely populated areas, domestic use exceeds other uses. Conservation efforts should include:
  - appropriate plantings and lawns
  - efficient water fixtures and appliance in times
  - efficient personal use
  - reuse of gray water
  - repair leaks and monitor water usage
  - appropriate pricing of water usage
- Innovative technologies can reduce water usage by industrial and manufacturing by as much as 25% to 30%.
- The public perception of water supply is critical to ensure an adequate water supply. The public needs to realize that certain types of water usage are not compatible with the available

water supply. For example, the citizens of Tucson, Arizona use water in a manner that is more compatible to its desert environment than do the citizens of nearby Phoenix, Arizona.

What is sustainable water use?

- Sustainable water use may be defined as the use of water resources by people that allows our society to develop and flourish into an indefinite future without degrading the various components of the hydrologic cycle or ecological systems that depend on it.
- Sustainable water use criteria include:
  - sufficient water volume to maintain human health and well-being
  - sufficient water volume to maintain ecosystems
  - maintaining minimum water quality standards
  - ensuring that water resources can be renewed
  - promoting water-efficient technology and practice
  - appropriate pricing of water resources

What is water management?

- Water management entails controlling the volume and time of water resources to achieve a predetermined objective (e.g., hydroelectric power generation, municipal water supply, flood control, or wildlife habitat).
- According to Luna Leopold, successful water management decisions are based on an understanding of geologic, geographic, and climatic factors, as well as economic, social, and political factors. Managers must realize that surface water and groundwater resources fluctuate with time and that advanced planning will be required to ensure sustainability.
- Water management for human use is sometimes incompatible with water needs of natural ecosystems.

What role do wetlands play in water management?

- Wetlands are ecosystems that are saturated with water for a portion of the year, have characteristic hydric soils, and wetland-adapted vegetation. These three components are used to define wetlands for regulatory purposes.
- There are several types of wetlands, some of which include: coastal and brackish marshes, freshwater marshes, forested swamps of floodplains and low-lying areas, peat bogs, isolated prairie potholes, and vernal pools.
- Wetlands have specific ecological functions, some of which are considered ecological values. Functions are ecological processes that occur in an ecosystem; if these benefit humans, they are considered values as well. Wetland functions vary according to wetland type.
- Wetland functions and values include but are not limited to:
  - *Function*: surface water storage; *value*: reduces downstream flooding
  - *Function*: groundwater recharge; *value*: resupplies aquifers for well water production
  - *Function*: wildlife nursery area; *value*: production of desirable wildlife, fish, or shellfish species
  - *Function*: uptake of nutrients, pollutants, and sediment trap; *value*: improves water quality
  - *Function*: buffer storm surge; *value*: protects upland from shoreline erosion

    - *Function*: carbon storage; *value*: source of organic carbon and retention of potentially harmful greenhouse gases
- Over 50% of the wetlands that were present in the United States 200 years ago have been drained or degraded for agricultural, industrial, urban, and residential development.
- Wetlands are considered "waters of the United States" and are regulated by federal Clean Water Act and state laws as such. In general, wetland regulations require that permits must be obtained before landowners undertake activities that may degrade wetlands. Considerable debate over wetland regulations and private property rights has ensued.
- Attempts to create, restore, or enhance wetlands in order to mitigate wetland losses elsewhere have not been entirely successful. The science and "art" of ecosystem restoration is still in a stage of development.

What effects do dams and reservoirs have on the environment?
- Dams and their reservoirs are constructed to manage water. The water may then be best used for a variety of purposes: municipal or agricultural water supply, hydroelectric power generation, flood control, improved navigation, or recreation.
- Dams and reservoirs have significant environmental impacts:
    - conversion of flowing water ecosystems into still water ecosystems
    - submergence of shoreline habitats
    - trapping of sediments in reservoirs behind dams and sediment starvation downstream from dams
    - water temperature alteration which affects spawning behavior of aquatic organisms
    - interference or blockage of migratory fish routes or damage to fish moving through hydropower turbines and sluiceways
    - alteration of water table and groundwater recharge and discharge patterns
- See the Case Study, *Removal of a Dam: A Maine Concern* to better understand the history and environmental consequences of river damming.

What effects do canals have on the environment?
- The excavation of canals is a water management technique used to either drain areas that are too wet for a desired use or to convey water to sites that are too dry for a desired use.
- Ecosystems and the organisms comprising ecosystems are thus strongly altered by canals. For example: the disease schistosomiasis, which is carried by snails living in canals associated with the High Dam at Aswan and Lake Nasser has become a human health problem in Egypt.

What is stream channelization?
- The deepening, widening, clearing, or lining of existing streams is called stream channelization. Streams are channelized to facilitate flood control, improve drainage, control erosion, and improve navigation (Figure 20.19).
- The environmental effects of channelization include:
    - alteration of stream habitat due to changes in stream morphology and hydrology
    - removal or alteration of streamside vegetation and habitat
    - removal of water storage capacity of streamside wetlands, thus more severe flooding downstream

    - elimination of wetland functions in adjacent wetlands
    - loss of aesthetic quality
- The Kissimmee River in central Florida was channelized in the 1960s in an unsuccessful attempt to control floodwaters. The U.S. Army Corps of Engineers is now attempting to restore the river to a condition that approximates its former morphology and hydrology (Figure 20.20).

How have humans affected flood patterns?

- Flooding is a natural and essential feature of many ecosystems.
- Along rivers and streams, floodwaters inundate floodplains transporting nutrients and sediments to the floodplain ecosystem.
- Floodplains are very important habitats for many plants and animals, and in many regions, floodplains are some of the last remaining tracts of undeveloped land.
- Periodically, flooding becomes a problem for humans when we build permanent structures, farm, or ranch on floodplains. Flooding is a natural phenomenon that becomes a human catastrophe.
- Efforts to control flooding have resulted in significant changes in the landscape. Dams and reservoirs have been constructed to control water releases, levees have been constructed to keep rivers within artificial banks, and diversion canals have been excavated to divert floodwaters from population centers.
- Flood control efforts may be successful locally, but often exacerbate flooding problems downstream.
- Urbanization creates impervious surfaces (e.g., pavement, roof tops) thus stormwater is not allowed to infiltrate into soil. Instead of slowly discharging into streams via groundwater, runoff is rapidly conveyed to streams; flash flooding results.
- Urban stormwater runoff can be highly polluted with grease, oil, solvents, heavy metals, and trash.
- The unfortunate flooding problems experienced along the Mississippi River in 1993 and the Red River in 1997 were, in part, a consequence of inappropriate land use on floodplains. See Case Study, *Bay of Pigs* at the beginning of Chapter 21 to better understand the consequences of inappropriate land use in the floodplains of the North Carolina Coastal Plain during the floods of 1999.

Ecology In Your Backyard

- What is your watershed address? Where does your water come from and where does it go?
- The US Geological Survey and the Environmental Protection Agency maintain websites with much information about the nations water supplies. Keyword search: USGS or EPA, *Surf your Watershed*. Or try the websites: http://h2o.usgs.gov/ or http://www.epa.gov/surf
- From a topographic map available in the library or that can be ordered at the USGS node of the National Geospatial Data Clearinghouse (http://nsdi.usgs.gov/nsdi/), find the drainage

basin in which you live and determine your watershed address. For example, the authors' watershed address in Greenville, NC is:

- Town Creek
    - Tar River
        - Pamlico River Estuary
            - Atlantic Ocean

- Where does your water supply come from? What river, reservoir, and/or aquifer supplies your water? What is the discharge volume in local rivers? Where is your waste water discharged?
- Trace the hierarchy of receiving waters from the point of discharge to the ocean.
- Contact your local city water authority and find out where you get and discharge water (name the aquifers or rivers involved).
- If you don't live where there is a municipal water supply and sewage, then find out where you get water from a well and discharge it to a septic tank.
- What federal , state, or local agencies are the principal regulatory agencies that manage your water supply?
- Botkin and Keller cite the daily per capita water usage in Tucson, Arizona, and Phoenix, Arizona as well as globally. From your local water supply authority, find out the per capita water usage in your area and the rates (billings) assessed.

Links In The Library

- Doppelet, B., M. Scurlock, C. Frissell, and J. Karr. 1993. *Entering the Watershed: A New Approach to Save Americas River Ecosystems*. Island Press, Washington. 462 pp.
- Heath, R. 1987. *Basic Ground-water Hydrology*. U.S. Geological Survey Water Supply Paper 2220. U.S. Government Printing Office, Washington.
- Leopold, L.B. 1994. *A View of the River*. Harvard University Press, Cambridge.
- McPhee, J. 1971. *Encounters with the Archdruid*. Farrar, Straus and Giroux, and McFarland, Walter and Ross, Toronto.
- van der Leeden, F., F.L. Troise, D.K. Todd.1990. *The Water Encyclopedia*. Second Edition. Lewis Publishers. 808pp. A sourcebook of statistical information concerning water resources worldwide, including climates, hydrology, surface and ground water, water use, water quality, water management, water resource agencies, legislation, and water-related environmental problems.

Ecotest

1. Which of the following is not a physical characteristics of water?
a. Water has a high surface tension.
b. Water has a high capacity to store heat.
c. Ice (solid water) has the highest density among water vapor, ice, and liquid water.
d. Water has the capacity to dissolve a great variety of compounds.
e. Transparency of water allows photosynthesis to occur in surface waters.

2. The residence time for water in the world's water supply is shortest for:
a. the oceans.
b. the atmosphere.
c. surface waters.
d. ground water.
e. ice caps and glaciers.

3. A model which accounts for the pathways (inputs and outputs) of water in a system is called a(n):
a. water budget.
b. aquifer test.
c. liquidation algorithm.
d. coliform test.
e. logistic model.

4. Where would one expect to find groundwater overdraft?
a. where recharge rates of the aquifer exceed rates of withdrawal
b. where recharge rates of the aquifer are less than rates of withdrawal
c. where recharge and withdrawal rates are in equilibrium
d. where the water table is very far below the surface

5. Use of water for irrigation of farmland is an example of:
a. off-stream use.
b. on-stream use.
c. domestic use.
d. multiple use.

6. The three factors that are used to determine the presence of a wetland include hydrology, vegetation type, and ____________.
a. soil type
b. topography
c. presence of animal tracks
d. pH

7. Which of the following is not a recognized wetland function?
a. high productivity
b. water storage
c. carbon storage
d. wildlife habitat
e. All of the above are functions of wetlands.

8. Environmental effects of dams include all of the following except:
a. loss of land.
b. increased sedimentation above the dam.
c. increased sedimentation below the dam.
d. downstream changes in hydrology.
e. alteration of habitat for fishes.

9. The straightening, deepening, widening, clearing, or lining of existing stream channels is called:
a. channnelization.
b. floodplain development.
c. accretion.
d. erosion.
e. subsidence.

10. According to Botkin and Keller's textbook, a water management plan that recognizes the natural variability of water supplies due to geologic, climatic, and geographic factors as well as the traditional social, economic, and political factors has been developed by:
a. Luna Leopold.
b. U.S. Army Corps of Engineers.
c. Arcadis, Geraghty & Miller.
d. the City of Phoenix, AZ.
e. the South Florida Water Management District.

11. _____ of the Earth's waters is in its oceans.
a. 97%
b. 80%
c. 21%
d. 70%
e. 50%

12. Today, it is estimated that total worldwide water usage is:

a. 1000 km^3/yr
b. 6000 km^3/yr
c. 600 km^3/yr
d. 10 km^3/yr

13. Of total precipitation, _______ enters groundwater or surface water supplies.
a. 3 %
b. 34 %
c. 66 %
d. 90 %

14. ____ of the water used in the United States is for agricultural purposes.
a. 15 %
b. 40 %
c. 65 %
d. 80 %

15. Underground soil or rock bodies that contain pore spaces or fractures, which may become filled with water, are called:
a. wells.
b. sinkholes.
c. aquifers.
d. hydroponics.

Chapter 21 - WATER POLLUTION AND TREATMENT

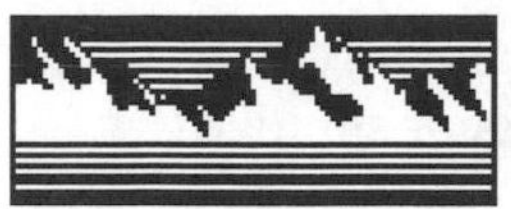

The Big Picture

By the 1960s, water pollution had become a major environmental and human health issue in many urban as well as rural areas of the United States. As a result of the near biological death of such places as Lake Erie and the Hudson River, citizens and legislators responded by promoting and enacting federal and state legislation aimed at reducing pollution inputs and cleaning up existing water pollution problems. Agencies have now been established to measure and monitor water quality to ensure that drinking water and water in streams, rivers, and estuaries meets basic standards for water quality. If waters are found to be out of compliance with these standards, legal actions may be undertaken to remediate the problems. Water quality parameters that are routinely monitored include biological oxygen demand (BOD), dissolved oxygen concentrations (DO), coliform bacteria, nutrient concentrations, oils, sediments, and an array of hazardous and toxic chemicals. Pollutants may enter waterways from specific locations called point sources, such as wastewater discharge pipes, or from diffuse locations, called nonpoint sources, such as agricultural and urban runoff. As a consequence of public involvement, legislation, and the application of new technologies to manage and treat wastewater, many areas that were severely polluted decades ago are now much improved. However, as human populations increase and relocate to new regions of the country water pollution problems often accompany these changes. In addition, new and complex toxic chemicals are being introduced into the environment and ultimately into water from industry, agriculture, runoff, and air pollution. Worldwide, water pollution and waterborne diseases are enormous environmental and human health problems; from densely populated India where fecal coliform bacteria concentrations are measured in millions of cells per 100 ml to the seemingly pristine lakes of central Canada where fish are contaminated with mercury.

Frequently Asked Questions

What are water pollution and water quality?

- **Water pollution** is the degradation of physical, chemical, or biological properties of water, beyond normal conditions. These properties comprise water quality.
- **Water quality** determines how water can be used. For example, if water is to be used as a supply of drinking water, it must be free of toxic chemicals and pathogens that cannot be removed through normal treatment processes. However, water that contains toxic chemicals and pathogens may be suitable for industrial processing.
- Water pollution may originate from any sector of society: urban, rural, industrial, agricultural, and military (Table 21.1).

What standards are used to measure water quality?

- In the United States, the Environmental Protection Agency (EPA) has established sets of **water quality standards** (threshold concentrations or conditions) for several common pollutants (Table 21.2). If these standards are exceeded, water may not be legally used for certain purposes. States must adopt these standards or develop more stringent standards. Standards differ according to water classification; drinking water standards are more stringent than standards for water in streams and rivers.
- **Water quality monitoring** (systematically sampling and analyzing water or wastewater) is usually accomplished by state environmental agencies. The EPA maintains a national database (STORET) of water quality data.
- A federal program to assess and monitor the nation's water quality is the **National Water Quality Assessment** (NAWQA) program, which attempts to standardize sampling regimes and techniques, and data analysis in major watersheds throughout the United States.

What categories of surface-water pollution are typically monitored?

Biochemical Oxygen Demand (BOD) and Dissolved Oxygen (DO)

- BOD and DO are not pollutants, but are critical aspects of water quality. The decomposing organic remains of aquatic plants and animals, or organic debris that is carried into waterways by natural processes (e.g., wind, runoff, flooding) forms the basis for decomposer food chains. However, excessive organic matter may also be a pollutant.
- Microbes utilize oxygen in decomposition processes. Oxygen usage can be measured and reported as BOD. Excessive organic matter and high decomposition rates may reduce the concentration of dissolved oxygen. As BOD increases, DO typically decreases.
- Species of fish and invertebrates that are intolerant of abnormally low dissolved oxygen concentrations cannot survive, and are replaced by more tolerant species.
- Sources of excessive organic matter include improperly functioning sewage treatment facilities, livestock operations, urban runoff, as well as the decomposing remains of algae and aquatic plants resulting from eutrophication.
- In flowing waters, water quality gradually improves downstream from the polluted site. Initially, there exists a *polluted zone* characterized by high BOD and reduced DO. Further downstream, in the *active decomposition zone*, microbial decomposition is at a maximum and DO is minimum. In the *recovery zone*, most of the organic matter has been consumed, microbial decomposition and BOD are reduced, and DO concentrations recover.
- Dissolved oxygen concentrations are also closely coupled with water temperature and movement. DO concentrations tend to be highest in cold and turbulent waters with low BOD. Conversely, warm, still waters with high BOD tend to have low DO concentrations.

Waterborne pathogenic microbes

- Waterborne diseases are responsible for the illness or death of millions of humans each year. Amoebic dysentery and cholera are a few of the pathogenic diseases that may be contracted from contaminated drinking water.
- In the more developed nations, drinking water supplies usually receive sufficient treatment to eliminate most pathogens. In the less developed nations, treatment of drinking water supplies is often inadequate and disease transmission is much more common.

- Because it is difficult to monitor water for specific pathogens, fecal coliform bacteria are used as indicators of potential water contamination. Fecal coliform bacteria reside in the gut of endotherms ("warm-blooded" animals, including humans); thus the presence of fecal coliform in water samples indicates the potential for pathogens to exist.
- The EPA has established standards for fecal coliform concentrations. Based on fecal coliform concentrations, waters are suitable for full contact activities (swimming and wading) if fecal coliform concentrations are less than 200 colonies/100 ml. Waters suitable for drinking may not contain any coliform bacteria, after treatment.

Nutrients

- Nitrogen and phosphorus are essential plant nutrients. In aquatic, estuarine, and marine ecosystems, the concentrations of these two nutrients may impose limitations on plant growth. In excessive quantities, nitrogen and phosphorous may stimulate the production of phytoplankton, filamentous algae, and aquatic macrophytes (submerged, floating, and emergent higher plants). This process is called **eutrophication** and usually results in a decline in water quality.
- Eutrophic waters are characterized by excessive nutrient input and subsequently increased plant production. Oligotrophic waters are nutrient poor.
- Nutrients enter waterways via runoff from agricultural fields, livestock operations, wastewater treatment facilities, as well as residential, urban, and industrial areas and golf courses.
- Nitrogen and phosphorus concentrations tend to be highest in regions where agriculture is the primary land use.
- Nitrogen, in the nitrate form, may cause human health problems if it infiltrates into drinking well water supplies. In some rural areas in the vicinity of intensive livestock operations, nitrate contamination has become a problem.

Oil

- Oil (i.e., crude oil, heating fuel, gasoline) is a water pollutant. Major oil spills capture news headlines and cause severe environmental degradation and risks to human health in localized areas.
- The immediate and most obvious effects of oil spills on the environment are the deaths of shellfish, finfish, shorebirds, waterfowl, and marine or aquatic mammals.
- After the spill has dissipated and clean-up efforts have been undertaken, the long-term effects of oil contamination may persist. Oil may persist in sediments and substrates, possibly disrupting food chains and damaging spawning and nursery areas for many years.
- The effects of oil spills are variable, dependent upon the characteristics of the oil (viscosity, degree of refinement, presence of toxic chemicals in the oil) and the physical conditions of the water (temperature, motion, depth).
- Areas that are most susceptible to oil spills are shipping terminals and routes, waters adjacent to refineries, and marinas.
- While oil spills are especially newsworthy, the daily and cumulative discharge of oil from boat motors, runoff from roadways, leaking storage tanks, and other relatively small discharges cause more widespread contamination.

Sediment

- By volume and mass, sediment (rock and mineral fragments, chiefly: sand, silt, clay).

- Two environmental problems stem from sediment pollution. Firstly, soil loss reduces the long-term soil productivity. Secondly, eroded sediment enters waterways, thus degrading water quality and filling streambeds, channels, lake bottoms, and harbors with sediment.
- Erosion and sedimentation problems are closely coupled with land use patterns. Cultivated cropland or areas denuded by intensive grazing or feedlots may have significant erosion problems. This is especially problematic in areas with sloping topography and high rainfall, or where soil conservation practices are not utilized.
- By contrast, permanent vegetative cover (i.e., forest or pasture) greatly reduces erosional losses. In general, water quality is much higher in watersheds dominated by forests compared to those dominated by agricultural production.
- Sedimentation emanates from urban areas as well. The construction of buildings and roadways may result in severe erosion, albeit most severe during the construction phase. Once the buildings or roads are completed, a vegetative cover may reduce but not eliminate erosion and sedimentation problems.
- Impervious surfaces, such as rooftops, paved parking lots and roads, concentrate the erosive force of water by diverting high volumes of runoff into small receiving areas.

What is acid mine drainage?

- Mining activities may result in the acidification of streams and other surface waters and groundwaters due to acid mine drainage.
- As surface runoff and groundwater move through mined areas or mine tailings, sulfuric acid is leached into these waters, lowering the pH to levels that may be toxic to most organisms.
- Streams in the coal mining regions of the Appalachian Mountains, Allegheny Plateau, and Rocky Mountains are especially polluted due to acid mine drainage. Copper, lead, zinc or other mines that liberate sulfuric acid as a by-product are also sources of acid mine drainage.

What are point and nonpoint sources of water pollution?

- Discharges of pollution from distinct and confined locations are considered **point sources** of pollution. For example, sewage or industrial waste discharges from pipes into waterways.
- Pollutant sources that are more diffuse and less confined than point sources are called **nonpoint sources**. For example, runoff from roadways and parking lots, lawns and other urban areas, agricultural fields and feedlots, mine sites, and some forest sites.

How can pollution from point and nonpoint sources be reduced?

- Point sources are more readily monitored, regulated, and controlled than nonpoint sources. One way that point source pollution is being reduced is through a federally- mandated permit system. In order for a factory, municipal wastewater facility, or other wastewater discharger to discharge into waterways, an **NPDES** (National Pollutant Discharge Elimination System) permit must be obtained. This program helps to ensure that pollutants in wastewater discharges do not exceed EPA standards.
- In general, there has been considerable improvement in water quality in the United States since the 1950s and 1960s. This has come about largely as a result of federal and state legislation and enforcement aimed at reducing point sources of pollutant discharges and implementing appropriate technologies.

- Nonpoint sources of pollution are much more difficult to regulate and treat. Federal and state environmental agencies have long recognized the problems associated with nonpoint source pollution, but the implementation of regulations and technologies is only now being attempted.
- Nonpoint source pollution is closely coupled with land use. Therefore, planning and implementing appropriate land use practices can reduce this type of pollution. This may entail such practices as: establishing permanent vegetative buffers between agricultural areas and waterways, maintaining highly erodable slopes in a vegetative cover, and creating catchment basins or artificial wetlands to collect and treat runoff.
- Perhaps more expensive and difficult to implement is the treatment of wastewater containing nonpoint source pollutants in wastewater treatment facilities. Many large municipalities will soon be required to treat urban runoff in a wastewater treatment facility before discharging to natural waters.
- Ecologists, planners, and many citizens are aware that if water quality is to be improved and maintained at a high level, water pollution issues will have to be addressed on a regional or watershed basis. State and federal agencies are developing and implementing watershed-water quality plans for many regions of the United States.

What are the common sources of groundwater pollution and how is this problem being addressed?

- Groundwater pollution is not widespread in the United States, but many local areas are experiencing problems with contamination. Approximately one-half of all people in the United States rely on groundwater for all or part of their drinking water supply, so it is vital to protect groundwater from contamination.
- Groundwater contamination often results from abandoned, illegal, or improperly functioning waste disposal sites. **Leachate** (contaminated water) infiltrates into groundwater supplies and moves in plumes through the aquifer.
- Leaking underground storage tanks (e.g., buried heating oil tanks at houses and gasoline tanks at gas stations) constitute a significant problem in the United States. Not only is there a risk of groundwater contamination, but the corrosive effects of petroleum could damage buried cables, water lines, and sewer lines.
- In the United States, an estimated 75% of the 175,000 known waste disposal sites may be producing plumes of hazardous chemicals that could migrate into groundwater.
- In some agricultural areas, pesticides and fertilizer may contaminate groundwater.
- In coastal areas, excessive groundwater withdrawals may lead to **saltwater intrusion** into freshwater aquifers, thus contaminating the water supply.
- The hazards presented by a specific contamination site depend upon the concentration and toxicity of the contaminants, exposure risk to humans and other organisms, as well as local geology and hydrology which affects dispersal capability of the contaminants.

How is wastewater treated?

- In most rural areas, there are no centralized wastewater treatment facilities. Most homes rely upon on-site, septic-tank disposal systems to treat domestic sewage. The effectiveness of septic-tank systems depends upon proper siting and maintenance.

- Septic-tanks function by separating solids from liquids, creating an environment that facilitates decomposition by microbes, stores organic matter, and clarifies liquids to be discharged. Once discharged into an absorption field, the treated wastewater is oxidized and filtered.
- In urban and industrial areas, wastewater is usually treated at centralized wastewater treatment facilities. Liquefied raw sewage is transported to a treatment facility via a network of sewer pipes. Treated wastewater is then discharged into receiving waters (river, lake, or ocean) or used for crop irrigation. The **Federal Water Pollution Control Act**, and amendments (**Clean Water Act**) stipulates that wastewater must be treated before it is discharged into natural waters.
- There are three categories of wastewater treatment methods: primary treatment, secondary treatment, and advanced (or tertiary) treatment.
    - ***Primary treatment*** entails separating particulate matter (sand, grit, plastic, paper), grease and oils from incoming sewage. In settling tanks, the application of alum or similar chemicals facilitates the precipitation of sludge, which will be collected and treated. Primary treatment reduces the pollutant volume of wastewater by 30% to 40%. In most municipalities, primary treatment is only the first step in wastewater treatment, to be followed by secondary treatment.
    - ***Secondary treatment*** utilizes activated sludge (which contains aerobic microbes) to decompose the organic sewage received from primary treatment. In aeration tanks, wastewater is mixed and kept aerated to promote decomposition. Next, the wastewater is transferred to sedimentation tanks where anaerobic microbes further degrade the accumulated sludge. Methane gas is a by-product. Finally, the wastewater is treated with chlorine (or other disinfectants) to kill bacteria or pathogens. The treated wastewater is then discharged into receiving waters or, in limited cases, used for irrigation. After secondary treatment, approximately 90% of the original pollutant volume is removed.
    - ***Advanced treatment*** entails using additional treatment steps to remove pollutants that still remain in wastewater after secondary treatment. These pollutants typically include nutrients (phosphate and nitrates), organic chemicals, and heavy metals. Specific technologies must be used to remove these pollutants; these technologies may be very expensive, but the expense may be necessary if the receiving waters are especially sensitive.
    - ***Chlorine treatment*** disinfects water by killing pathogens before wastewater is returned to receiving waters and is often used as part of secondary and advanced treatment. But chlorine, especially at high concentrations is hazardous to aquatic life as well as workers at treatment facilities.

Can wastewater be recycled and reused?

- Treated wastewater can be recycled but precautions must be taken to avoid the transmission of pathogens or contaminants.
- Under the proper site and usage conditions, treated wastewater may be used to irrigate crops, lawn, and golf courses, and to recharge aquifers.

- Wetlands have been used to receive discharged wastewater, though this is largely experimental. Wetland plants are capable of absorbing excessive nutrients and contaminants, thus removing them from water supplies. The widespread use of natural wetlands for wastewater recycling is unlikely; however, the use of artificially constructed wetland may be more promising.

How safe is drinking water?

- In the United States, drinking water is obtained from groundwater and surface water supplies. Some groundwater is of high quality and may not need treatment prior to consumption. However, most surface water requires treatment to remove pathogens or other contaminants.
- Chlorination is a widely used method to eliminate most pathogens, however, the chlorine added to water may result in adverse taste or may present a health risk itself.
- Federal and state drinking water standards are designed to keep contaminant concentrations to a minimum. However, health complications arising from long-term exposure to low levels of contamination are poorly understood.

What laws and legislation have been enacted to control water pollution and safeguard water quality?

- The **Federal Water Pollution Control Act** (commonly referred to as the Clean Water Act) is the most comprehensive legislation targeting water pollution control. This Act includes important legislation such as requiring permits to discharge wastewater into waterways (NPDES permits) and legislation that provides protection for the nation's wetlands (Sections 404 and 401).
- The **Safe Drinking Water Act** empowers the EPA to establish drinking water standards.
- The **Water Quality Act** targets nonpoint source pollution.
- See Table 21.4 for information on other water pollution legislation.

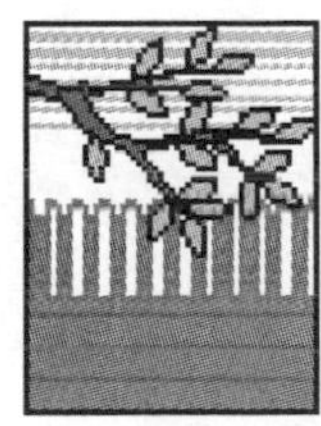

Ecology In Your Backyard

- Do you know the source of your drinking water? Clean water is absolutely basic to good health, however too often we do not know the source of our water or the fate of our wastewater.
- Arrange a visit to the water treatment facility and the wastewater treatment facility nearest you. Often, tours can be arranged to provide you with an overview of facility operation and laboratories.
- Contact your state fish and game or wildlife office to find out if advisories have been issued to limit or avoid human consumption of fish taken from local waters. Long-lived fish, bottom-feeding fish, or species at the top of aquatic food chains tend to bioaccumulate toxins.

Links In The Library

- Stednick, J.D. *Wildland Water Quality Sampling and Analysis*. 1991. San Diego: Academic Press, Inc. A concise review of general chemistry and water quality sampling techniques and standard procedures is presented prior to a discussion of typically measured water quality parameters. The background information and a table of typical values about each parameter is especially useful.
- Van der Leeden, F. *The Water Encyclopedia* 1990. Chelsea, Michigan: Lewis Publishers, Inc. A compendium of water-related information in tabular formats. Chapters 6 and 7 pertain to water quality and environmental issues.
- Terrene Institute. *Nonpoint Source: News-Notes*. This is a bimonthly newsletter published under a cooperative agreement with the U.S. EPA. Water-related environmental issues are highlighted.

Ecotest

1. What are EPA water quality standards?
a. a listing of all pollutants detected in water samples
b. federally certified water quality monitoring stations
c. a description of optimal water quality conditions for specific watersheds
d. threshold concentrations for water pollutants

2. All of the following are examples of nonpoint sources of pollution except:
a. agricultural runoff into drainage ditches.
b. aerosol deposition downwind from industrial centers.
c. discharge from a sewage treatment facility.
d. stormwater runoff from urban streets.

3. If nutrients are artificially supplied to aquatic and estuarine ecosystems, excessive plant growth may result and water quality will decline. This form of pollution is called _____.
a. enrichment
b. eutrophication
c. nutrient limitation
d. enhancement

4. Which of these statements about biochemical oxygen demand (BOD) and dissolved oxygen (DO) in streams is false?
a. BOD is highest and DO is lowest immediately upstream from sewage discharges.
b. As BOD increases, DO decreases.
c. Waters with abundant dead organic matter tend to have high BOD.
d. In the polluted zone and recovery zone of a stream, aquatic species that are adapted to low DO are more apt to survive.

5. Nitrogen and phosphorus concentrations in streams tend to be highest in streams located in watershed dominated by which land use category?
a. forest
b. urban
c. agricultural
d. industrial

6. An estimated ____ % of the 175,000 known waste disposal sites in the United States may be producing hazardous chemical plumes that could migrate into groundwater.
a. 85
b. 75
c. 55
d. 33

7. Waters are considered too polluted for swimming if fecal coliform concentrations are in excess of ____ colonies per 100 milliliters.
a. 50
b. 75
c. 100
d. 200

8. Which statement about the operation of wastewater treatment facilities is false?
a. Chlorine is injected into sludge to eliminate bacterial growth.
b. Secondary wastewater treatment does not eliminate nitrogen, phosphorus, and many toxic materials.
c. An aerobic environment is maintained in the aeration tank, whereas, an anaerobic environment is maintained in the sludge digester.
d. Grit chambers and sedimentation tanks remove approximately 30% to 40% of the pollutant volume from wastewater in the primary treatment phase.

9. Which statement about federal water legislation is false?
a. The Water Quality Act (1987) targets point sources of pollution.
b. The Federal Water Pollution Control Act was enacted in 1972 during the Nixon Administration.
c. The Comprehensive Environmental Response, Compensation, and Liability Act provides funds for clean up at hazardous waste disposal sites, thus reducing groundwater pollution.
d. Contaminant standards were established in the Federal Safe Drinking Water Act.

10. In coastal areas that rely on groundwater supplies, intensive groundwater removal (well pumping) may draw salt water into the freshwater aquifer and contaminate groundwater supplies. This is called ______.
a. salinization
b. salt leaching
c. salt water intrusion
d. deep well injection

11. Which method of cleaning polluted water is not mentioned in the text above?
a. Adding lime to streams where the pH is too low.
b. Creating artificial wetlands to remove nutrients.
c. Using wastewater to irrigate crops.
d. Boiling or distilling drinking water.

12. Impervious surfaces, such as paved parking lots, tend to _______ the erosive force of water in adjacent areas.
a. dissipate
b. concentrate
c. interrupt
d. have no effect on

13. Primary treatment of wastewater entails:
a. separating out the larger particulate matter from the incoming wastewater.
b. allowing microbes to decompose organic sewage.
c. heavy metal removal.
d. treatment with chlorine to kill bacteria and pathogens.

14. What fraction of all people in the United States rely on groundwater for all or part of their drinking water supply?
a. 1/2
b. 1/3
c. 1/4
d. 1/5

15. In the active decomposition zone of an effluent-receiving stream,
a. microbial decomposition and BOD are at their highest, DO is at its lowest.
b. all microbial activity has been completed.
c. water quality has improved since the water has flown down from the area of pollution.
d. BOD is minimal.

Chapter 22 - THE ATMOSPHERE, CLIMATE, AND GLOBAL WARMING

The Big Picture

The Earth occupies a unique situation in the solar system. Due to its position 149 million km from the sun, physical conditions on Earth are conducive to life. At this position, the temperature range at the Earth's surface allows water to exist in solid, liquid, and vapor states and atmospheric gases to exist in an appropriate mix and concentration. The solar radiation that reaches the Earth is either reflected by the atmosphere or temporarily absorbed by the oceans, waters, soils, or biota. Ultimately, all solar energy that reaches the Earth is reradiated to space; in the meantime however, solar radiation drives weather, climate, and ecosystems. The amount and timing of solar radiation that is absorbed is unevenly distributed. Equatorial regions receive more total solar radiation annually than temperate or polar regions. The excess of energy is redistributed poleward by oceanic and atmospheric circulation. This redistribution of energy and the insulating properties of the atmosphere prevent the extreme variations in temperature and moisture that inhibit most life forms. However, climatic differences do exist (mainly due to temperature and moisture variations) and are the most important factors determining the distribution and abundance of species. Humans are the most dominant species on Earth. Due to technological advances and sheer population numbers, most scientists believe that humans are beginning to alter the planet's climate. Since the beginning of the industrial revolution, humans have become increasingly reliant upon fossil fuel combustion as an energy source. The greenhouse gases produced by fossil fuel combustion and from other anthropogenic sources coupled with deforestation and changing land use patterns are probably responsible for recent increases in average annual global temperature. The consequences of global warming are enormous. Changing temperature and moisture patterns will have a significant effect on natural and agricultural ecosystems, sea level, and human population distribution. There is considerable and unresolved debate as to the causes, outcomes, and solutions to the global warming situation.

Frequently Asked Questions

What is the Earth's atmosphere composed of?

- Seventy-eight percent of the atmosphere is free nitrogen (N_2), an inert gas.
- Twenty-one percent of the atmosphere is free oxygen (O_2).
- The remaining 1% is most inert argon (Ar) (0.9%), carbon dioxide(CO_2) (0.03%), and trace amounts of other gases and particulates.
- The relative percentage of these gases remains constant throughout the atmosphere, but the density of atmospheric gas molecules decreases with altitude.
- Water vapor constitutes from 1% to 4% of the troposphere, the layer nearest the Earth's surface.

- Atmospheric pressure is a measure of gas density. For example, atmospheric pressure at sea level is approximately 10^5 N/m^2 or 14.7 lb/in.^{2}
- Life on Earth is dependent upon three basic physical conditions of the atmosphere: gases at the appropriate ratios and atmospheric pressure, atmospheric temperature within an appropriate range (Figure 21.1), and relative humidity.

What are atmospheric conditions like in the troposphere and the stratosphere?

- The **troposphere** is the lowest stratum of the atmosphere ranging from the Earth's surface to 10-12 km. The troposphere is the only stratum with an abundance of life, water, and weather. Air temperature and pressure decrease with increasing altitude (Figure 21.2).
- The **stratosphere** extends from the tropopause (~11 km) to 50 km in altitude; temperature is a relatively constant -55° C throughout. The stratospheric ozone layer between 20 km and 25 km protects the Earth from harmful exposure to ultraviolet radiation.

What are the key properties of the atmospheric that influence weather and climate?

- **Air pressure**- air moves from areas of relatively high pressure to relatively low pressure.
- **Air temperature**- temperature is influenced mainly by the intensity and duration of solar insolation but local conditions influence temperature as well. These include cloud cover, vegetative cover, proximity to water, altitude, and air masses.
- **Relative humidity**- relative humidity, expressed as a percentage, is a measure of the actual amount of water vapor in the atmosphere relative to the potential amount of water the atmosphere can contain. It is dependent upon three basic factors: air temperature, air pressure, and water availability. An air mass containing the maximum amount of moisture possible given these three factors is referred to as *saturated*, and has a relative humidity of 100%. A rapid drop in air temperature, such as typically occurs at dusk, lowers the saturation point and water vapor condenses as dew.
- **Precipitation**- when relatively humidity reaches 100% water condenses (especially on dust particles) and precipitates out of the atmosphere. The water cycle is driven by the processes of evaporation and condensation.

What causes global air circulation and wind?

- Air moves from areas of relatively high air pressure to areas of relatively low air pressure. The stronger the gradient between high pressure air masses and low pressure air masses, the higher the wind speed.
- Differential heating and cooling of the Earth's surface and atmosphere creates differences in air pressure.
- At the equator, air near the surface is warmed, causing it to rise and creating a low pressure near the surface. To compensate for the low pressure area that is created, air masses move horizontally across the Earth's surface toward the equator from the north and south. In the upper troposphere above the equator, rising air cools and moves poleward then descends between 25° and 30° latitude. Similar circulating cells become established in the temperate zones and polar zones.

- Additional directionality is given to global wind patterns as a result of the **Coriolis Effect**. The Earth's rotational speed at the equator is faster than at higher latitudes. This tends to deflect surface winds to the right in the northern hemisphere and to the left in the southern hemisphere.
- The global wind belts are the:
  - **doldrums**: equatorial, low pressure, ascending air, little horizontal air movement
  - **horse latitudes**: ~30° N and S, high pressure, descending air, little horizontal air movement
  - **trade winds**: tropical zone, horizontal movement in an easterly direction
  - **westerlies**: temperate zones, horizontal movement in a westerly direction
  - **polar easterlies**: polar zones, horizontal movement in an easterly direction

What four processes are responsible for removing human-induced particles from the atmosphere?

- sedimentation- settling out of heavier particulates
- rain out- physical and chemical flushing of atmosphere
- oxidation- combination of aerosol chemicals with oxygen
- photodissection- breakdown of chemical by sunlight

What are weather and climate?

- **Weather** is short-term (daily to weekly) atmospheric conditions.
- **Climate** consists of long-term, characteristic weather patterns (Figure 22.5).
- Neither weather nor climate are static; both change over time, albeit climatic changes occur at a much slower rate.
- Precipitation and temperature are the two factors that most strongly influence climate.
- Climate, in turn, exerts a strong influence on the distribution and abundance of the Earth's biota.
- While biogeography is determined largely by global climate, individual organisms are influenced by microclimate (very localized atmospheric conditions).

What is urban microclimate?

- Cities alter local weather and climate.
- Tall buildings, concrete and asphalt surfaces, few trees, and fossil fuel emissions exert influences on the urban microclimate.
- Compared to adjacent rural areas, cities are generally warmer, dustier, and cloudier, with a lower relative humidity, higher precipitation, less solar radiation, and calmer winds.
- One effect of stagnant, polluted air is the development of urban heat domes and dust domes over cities (see Chapter 26).

Has global climate changed over time?

- There have been numerous major changes in global climate throughout geologic history of the Earth (4-5 billion years). During the Pleistocene (2 million years BP to 10,000 years BP) four major glaciations occurred (the most recent of which formed what is now Long Island, NY). Figure 22.6 illustrates changes in global temperature during the past 1.0 million years.

- **Glaciations** (also called glacial events or ice ages) are separated by **interglaciations**. It is unknown if the ice ages have ended or we are simply in an interglaciation.
- On a shorter time frame, periods of warming and cooling, and wet and drought have occurred. These cycles have had a strong influence on human settlement and history.
- In the last 100 years, the global mean annual temperature has increased approximately 0.5° to 0.7°C.
- The last two decades have been the warmest on record since global temperatures have been monitored (140 years). However, there is not sufficient evident to indicate whether this increase is solely the result of human influences on the global climate or part of a natural cycle of climatic change, or both. None-the-less, there is ample evidence that humans have changed the composition of the atmosphere (increasing greenhouse gases) and disrupted global biogeochemical cycling.

What are global warming and the greenhouse effect?

- **Global warming** is a natural or human-induced increase in average global temperature.
- The temperature of the Earth near its surface is dependent upon the amount of solar energy that is: 1) received, 2) reflected by the atmosphere and surface, 3) retained by the atmosphere, and 4) is transferred in evaporation and condensation.
- The **greenhouse effect** is a natural phenomenon. Atmospheric gases, primarily water vapor, trap radiant heat in the lower atmosphere. The greenhouse effect moderates daily and seasonal temperature fluctuations.
- Water vapor is a natural greenhouse gas; of greater concern for global warming are anthropogenic inputs of other greenhouse gases.

What is the relative contribution of the major anthropogenic greenhouse gases to the greenhouse effect?

- Water vapor has the greatest influence on the greenhouse effect but it is not an anthropogenic greenhouse gas although the concentration of water vapor in the atmosphere may be indirectly influenced by human activities.
- The contributions of anthropogenic greenhouse gases are listed in the table below and Table 22.1 in text.

Greenhouse Gas	Relative Contribution
CO_2	50% - 60%
CFC	15% - 25%
CH_4	12% - 20%
N_2O	5%

What is the status of atmospheric CO_2?

- Carbon dioxide (CO_2) concentrations can be measured in glacial ice to determine long term trends.
- Based on this evidence, prior to the industrial revolution in the 1700s, background concentrations of atmospheric CO_2 were 200-300 ppm.
- In 1860 CO_2 concentrations were 280 ppm.

- Today, CO_2 concentrations are ~400 ppm.
- Rate of CO_2 increase is 0.5% per year.
- Total carbon emissions (from CO_2, CH_4, other hydrocarbons, particluates, etc.) have increased 4.3% annually.
- The discrepancy between total carbon emissions and CO_2 concentration suggests that an unknown carbon sink (probably the oceans and/or terrestrial plants) is accumulating some of the excess carbon.

What is the status of atmospheric CFCs?

- **Chloroflourocarbons** (CFCs) are increasing at approximately 5% per year.
- CFCs are chemicals that have been or are used as aerosol spray propellants and refrigerants.
- CFC molecules are especially significant contributors to the greenhouse effect because they absorb wavelengths that would otherwise exit the atmosphere via an "atmospheric window."
- CFCs have a long residence time in the atmosphere.

What is the status of atmospheric CH_4?

- Methane (CH_4) is increasing 1% per year.
- Sources of CH_4 are termites, wetlands, cattle, rice cultivation, and biomass and fossil fuel combustion.

What is the status of atmospheric N_2O?

- Nitrous oxide (N_2O) is increasing as much as 5% per year.
- Sources include agricultural fertilizers and fossil fuel combustion.
- N_2O also has a long residence time in the atmosphere.

Is global warming really occurring?

- Because global atmospheric conditions apparently undergo cycles of warming and cooling, and wet and dry, it is difficult to attribute atmospheric conditions of any one year (or decade) to global climate change. Consequently, there is considerable debate among and between scientists, legislators, economists, and consumers about the causes and consequences of global warming.
- However, certain facts are apparent:
  - The greenhouse effect is a well-understood atmospheric phenomenon.
  - Anthropogenic processes increase the concentrations of greenhouse gases.
  - A correlation between atmospheric CO2 concentrations and global temperature has been established.

What can be done to better understand climate change?

- Evaluate geologic record and climatological data to understand change from long-term perspective
- Monitor and document recent and on-going changes
- Use atmospheric models to predict consequences
- Provide information to decision-makers

What scenarios and models of global warming have been developed?

- Most scientists expect that average annual temperature will be 2° - 4° C higher by mid 21st century.
- If global warming occurs, positive and negative feedback loops could increase or decrease global temperature.
- **Positive feedback** scenarios are as follows:
  - increased global warming increases evaporation, increased water vapor (a major greenhouse gas) causes an increase in the greenhouse effect and increased global warming.
  - increased global warming increases permafrost melting, thus releasing methane (a greenhouse gas) causing additional warming.
  - increased global warming initiates increased fossil fuel use for cooling, thus the emission of more greenhouse gases.
- **Negative feedback** scenarios are as follows:
  - increased CO_2 increases oceanic algae or terrestrial plant CO_2 uptake, the total amount of CO_2 would decrease and the greenhouse effect would diminish.
  - increased CO_2 increases evaporation rates causing increased cloud cover, less solar radiation reaches the surface, thus cooling the lower atmosphere.
- **Global Circulation Models** (GCMs) use atmospheric temperature, relative humidity, and wind conditions to predict the regional effects of global warming.
- GCM models are crude but indicate that by the year 2030 an increase of 1° - 2° C rise will occur, but this may be higher in polar regions.
  - In central North America, a predicted increase in winter precipitation and decrease in summer rains will result in decreased soil moisture during the growing season and decreased crop yields.
  - In Canada and Russia, the growing season may be increased but due to poor soil, crop yields may not result.
- Sea level will rise with global warming. (Actually, sea level is already rising at a rate of 2.5 - 3.0 mm per year.)
  - Two factors contribute to sea level rise, thermal expansion and the melting of glacial and polar ice.
  - Regional geology also determines the net effect of sea level rise; areas experiencing geologic subsidence will be more rapidly inundated, areas being uplifted may not be directly affected.
  - Sea level rise will cause a landward migration of coastal and maritime ecosystems and saltwater intrusion into aquifers and surface waters.
  - The economic costs of attempting to prevent inundation by building seawalls or abandoning coastal areas will be enormous.
- Biological systems would be affected by climate change. High latitude ecosystems, such as arctic tundra, may be compressed or eliminated. The distribution of species would change, the range of some species would expand while others shrink; interspecific interactions would change as well. Pathogenic organisms now confined to tropical regions may expand into the higher latitudes.

What other forces in the global warming question need to be better understood?

- **Sunspot cycles** is correlated with global temperature trends, but correlation does not necessarily mean cause-and-effect. Increased sunspot cycle length correlates with decreased global temperature and vice versa.
- **Aerosols** (particles <10um) provide condensation sites for water vapor. Increased cloud cover can result; this would increase reflectance of solar radiation and cause a cooling trend. Volcanic eruptions increase aerosol concentrations.
- **El Nino** reduces the amount of CO_2 outgased from the ocean to the atmosphere, thus altering the global carbon cycle.
- **Methane release** from the seafloor could occur as sea level changes. Methane is a greenhouse gas.
- **Anthropogenic forces** are most certainly a part of the problem but we could mitigate the magnitude of the problem by making the appropriate energy and land use choices.

What do we need to know about global warming in order to make the best possible decisions?

- There are many uncertainties surrounding our understanding of global climate and human interaction, chiefly:
  - we do not know with certainty if the global temperature increase that has occurred during the past 100 to 150 years will continue,
  - we do not know with certainty if the continued combustion of fossil fuels and increased atmospheric CO_2 concentrations will cause global warming to continue,
  - we do not know with certainty if policy decisions and economic sacrifices to curb emissions of greenhouse gases will have a beneficial effect.
- Given these uncertainties, we have two basic options.
  - Attempt to mitigate the severity of global warming by reducing greenhouse gas emissions. ("erring on the side of caution")
  - Accept that change will occur and adapt to the new conditions.

What is being done on an international level to address climate change?

- The 1992 Earth Summit in Rio de Janeiro was one of the first attempts to reach international consensus about climate change and ways to mitigate the consequences. A non-binding agreement was reached among some participating nations to reduce emissions of CO^2. But the US and others could not agree on the appropriate CO^2 levels.
- The Koyoto (Japan) Protocols in 1997 attempted to establish legally binding reductions but once again there was disagreement as to the appropriate levels and ways to mitigate climate change. For example, the US proposed that increasing reforestation efforts would offset CO^2 emissions while allowing CO^2 generating industrial and transportation activities to continue unabated.
- Such international forums are invaluable since climate change must be addressed on a global scale but the forums also highlight the highly variable economic and environmental perspectives of the global community.

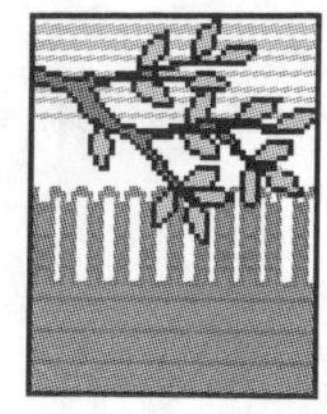

Ecology In Your Backyard

- Differences in microclimate can be extreme on hot summer days. Measure and compare air temperature at an elevation of 1 m and at 10 cm at a variety of different urban and rural sites. Readings should be taken at approximately the same time of day for valid comparisons. An interesting comparison would be an asphalt parking lot and a shaded park or a field compared to a forest on a sunny summer day.
- If you live in a low-lying coastal area, what effects might be seen if sea level rises as much as 30 cm (1 foot) over the next century?

Links In The Library

- Barron, E.J. 1995. Climate Models: How reliable are their predictions? *Consequences* 1 (3): 16-27.
- Kareiva, P.M., J.G. Kingsolver, and R.B. Huey. 1993. *Biotic Interactions and Global Climate Change*. Sinauer Associates, Inc., Sunderland, Massachusetts. 559 pp.
- Karl, T. R., R. W. Knight, D. R. Easterling, and R. G. Quayle. 1995. Trends in U. S. climate during the twentieth century. *Consequences* 1 (1): 2- 12.
- Lean, J. and D. Rind. 1996. The sun and climate. *Consequences* 2 (1): 26-36.

Ecotest

1. In the troposphere, water vapor concentration is variable but _____ is the usual range.

a. 1% - 4%
b. 5% - 7%
c. 9% - 11%
d. 10% - 13%

2. Most of the coterminous United States lies within the ______ wind belt.

a. northeast trades
b. westerlies
c. easterlies
d. southeast trades

3. Which statement about the troposphere is false?
a. The troposphere contains almost all of the water vapor in the atmosphere.
b. The troposphere contains almost all of the terrestrial life on the planet.
c. The troposphere is characterized by changeable weather.
d. The troposphere contains the "ozone layer".

4. Highly localized climatic factors that individual organisms are exposed to is called ______.
a. habitat
b. life zone
c. microclimate
d. range

5. According to the Council on Environmental Quality and the Department of State, all of the following statements about urban compared to rural climatic conditions are true except:
a. annual mean urban temperatures are 0.5° -1.0° C higher in urban areas.
b. annual mean relative humidity is 6% lower in urban areas.
c. annual mean wind speed is increased by 20-30% in urban areas.
d. total radiation on horizontal surfaces is decreased by 15-20% in urban areas.

6. Natural "greenhouse" conditions are largely a consequence of the heat trapping by ______, which is responsible for about 85% of the natural greenhouse effect.
a. CO_2
b. CH_4
c. H_2O (vapor)
d. O_2

7. This atmospheric gas resulting from human-related activities is responsible for 50-60% of the anthropogenic "greenhouse" conditions affecting global warming.
a. CO_2
b. CH_4
c. H_2O (vapor)
d. O_2

8. All of the following trends have been substantiated (using acceptable scientific methods) about global climate change except:
a. atmospheric CO_2 concentrations increased from 280 ppm to 880 ppm in the past 50 years.
b. global mean annual temperature has increased 0.5 to 0.7°C in the past 200 years.
c. the decade from 1986-1995 was the warmest decade on record (dating back 130 years)..
d. sea level is rising at a rate of 2.5 to 3.0 mm per year.

9. Which of the following global climate change scenarios is an example of a negative feedback loop?

a. Increased CO_2 concentrations cause an increase in oceanic algae or terrestrial plant CO_2 uptake, the total amount of atmospheric CO_2 would decrease and the greenhouse effect would diminish.
b. Increased global warming increases evaporation, increased water vapor (a major greenhouse gas) causes an increase in the greenhouse effect and increased global warming.
c. Increased global warming increases permafrost melting thus releasing methane (a greenhouse gas) causing additional warming..
d. Increased global warming initiates increased fossil fuel use for cooling, thus the emission of more greenhouse gases.

10. According to your author, in addition to emissions of anthropogenic greenhouse gases, each of the following processes may contribute to the global climate change, except:
a. sunspot cycles.
b. tectonic shifts in crustal plates.
c. El Nino.
d. aerosol inputs from volcanic eruptions and other sources.

11. Ninety-nine percent of the atmosphere is comprised of what two gases?
a. free nitrogen gas and carbon dioxide
b. methane and carbon dioxide
c. oxygen and free nitrogen gas
d. carbon dioxide and oxygen

12. The actual amount of water vapor in the atmosphere relative to the potential amount of water the atmosphere can contain, expressed as a percentage, is called:
a. dew point.
b. saturation index.
c. relative humidity.
d. atmospheric pressure.

13. The short-term (daily to weekly). trend in atmospheric conditions is known as the:
a. climate.
b. relative humidity.
c. microclimate.
d. weather.
e. biogeography.

14. All of the following directly influence urban microclimate except:
a. paved surfaces.
b. area of tree coverage.
c. fossil fuel emissions.
d. mass transit systems.

15. The Coriolis effect gives added directionality to global wind patterns. This tends to deflect surface winds to the _________ in the northern hemisphere, and to the _________ in the southern hemisphere, relative to the equator.
a. right; right
b. right; left
c. left; right
d. left; left

Chapter 23 - AIR POLLUTION

The Big Picture

Until the second half of the twentieth century, air pollution and air quality were not considered a global concern. Although, locally and regionally, many heavily industrialized areas experienced poor air quality and human health was at risk. Now, however, air pollution and air quality are of global concern, as well as local and regional. The effects of air pollution are manifest far downwind from the sources of pollution and the cumulative effects of pollutant discharges have begun to affect global climate. Many of the air pollution problems we are experiencing today are historically linked to our dependence on the combustion of fossil fuels to satisfy domestic, industrial, and automotive energy demands. In the developed nations, improved and costly technologies and pollution regulations have decreased the units of air pollution generated per person, but problems still exist because populations are large. Urban dwellers who are exposed to high concentrations of pollutants experience higher mortality rates than people living in less polluted areas. Burgeoning cities in developing nations typically do not have adequate pollution control regulations and cannot afford the expensive technologies to control air pollution. In the developed and developing nations, air pollution problems extend far beyond the urban and industrial areas. Acid rain, ozone depletion, and global warming are international air pollution issues.

Frequently Asked Questions

When did air pollution become an environmental issue?

- Air pollution has been a localized problem since humans began using fire. Poor air quality in confined areas most certainly contributed to lung and respiratory diseases.
- Air pollution became a more regional problem in cities of the eighteenth century industrial revolution.
- Air pollution events in Donora, Pennsylvania (1948) and London, England (1952) caused numerous deaths and initiated the passage of regulations aimed at controlling and reducing air pollution. Though regulations have certainly attenuated air pollution, problems persist.
- Since the proliferation of automobiles, air pollution in cities such as Los Angeles, Mexico City, and Athens has become a severe environmental and human health problem. Beijing does not have as many automobiles but severe air pollution problems stem from its reliance on coal for heating and electrical production.
- Now air pollution is much more widespread. Not only are cities prone to air pollution problems, but rural areas, especially those downwind from industrial centers, power plants, and concentrated livestock operations are experiencing health and environmental problems.

What are the main sources of air pollution?

- Air pollutants arise from **natural sources** and **anthropogenic (man-made) sources** (Table 23.2).
- **Natural emissions** include volcanic and geothermal eruptions, decaying matter in wetlands, atmospheric events, dust storms, and wildfires.
- For most forms of air pollution, natural sources contribute more emissions than anthropogenic sources (with the exception of sulfur oxides and nitrogen oxide).
- The global ecosystem is adapted to natural pollutant levels; additional inputs by humans may disrupt the ability of the ecosystem to maintain homeostasis.
- **Anthropogenic sources** are categorized as either stationary sources or mobile sources.
- **Stationary sources** may be point sources (such as power plants), fugitive sources (such as dirt roads and construction sites), or area sources (such as urban or agricultural areas that generate air pollutants). **Mobile sources** include automobiles, trains, aircraft, and ships.

What are some of the general effects of air pollution on the environment and human health?

- Air pollution is aesthetically and economically undesirable because it reduces visibility, creates unpleasant odors, and damages statuary and other artificial structures.
- Air pollution damages plant leaves, interfering with photosynthesis, and pollutes the soil, causing root damage. Consequently, crop yields can be reduced and natural plant communities are altered.
- Animals may experience many of the same respiratory problems that humans experience. The apparent decline in amphibians that are especially sensitive to environmental contaminants may be attributable to air pollution.
- The effects of air pollution on humans depend upon concentration and duration of exposure to pollutants and individual susceptibility. Some effects include toxic poisoning, cancer, birth defects, eye irritations, irritation of the respiratory system, increased susceptibility to viral infections (causing bronchitis and pneumonia) and heart disease, and aggravation of chronic asthma and emphysema (Figure 23.6).
- The combined effects of multiple contaminants may have synergistically adverse health consequences. Urban dwellers are especially susceptible to multiple exposures.
- An estimated 150 million people live in areas of the United States where air pollution is a health risk. Air pollution is a contributing factor to as many as 120,000 deaths per year and the annual health care costs stemming from exposure to air pollution are an estimated $50 billion.

What are the major air pollutants?

- Air pollutants are either primary pollutants or secondary pollutants.
- **Primary pollutants**, such as particulates, sulfur dioxide, carbon monoxide, nitrogen oxides, and hydrocarbons, are emitted directly into the air.
- **Secondary pollutants** are pollutants that form when primary pollutants undergo chemical reactions in the atmosphere; these include tropospheric ozone and sulfuric acid in precipitation.

- **Sulfur dioxide** (SO_2) is a product of fossil fuel combustion (especially combustion of low-grade coal) and industrial processing. In the atmosphere SO_2 is converted to fine particulate sulfate (SO_4). SO_2 and SO_4 can damage plant and animal tissue. SO_2 and SO_4 are precursors of acid precipitation.
- **Nitrogen oxides** (primarily NO and NO_2) are products of fossil fuel combustion in automobiles and power plants. Nitrogen oxides are a major component of smog and are precursors to acid precipitation. Nitrogen oxides damage plant tissue and are respiratory tract irritants in animals.
- **Carbon monoxide** (CO) is a colorless, odorless gas that is readily absorbed by blood hemoglobin. If high concentrations of CO are present, oxygen uptake by aerobic animals may be impaired. CO poisoning is especially dangerous to fetuses and to persons with heart disease, anemia, or respiratory disease.
- **Ozone and other photochemical oxidants** such as peroxyacyl nitrates (PANs) are formed when nitrogen dioxide interacts with sunlight. Nitrogen dioxide emissions are from automobiles, fossil fuel combustion, and industrial processing. Ozone in the stratosphere provides an import barrier to harmful ultraviolet radiation, however, high concentrations of ozone in the troposphere can damage plant leaf tissue and the respiratory tracts of animals. Ozone also damages man-made materials such as rubber, paint, and textiles.
- **Volatile Organic Compounds (VOCs)** are hydrocarbons such as methane (CH_4), butane (C_4H_{10}), and propane (C_3H_8) are components of smog and in high concentration may cause respiratory problems and other serious complications in animals. Most hydrocarbons (>80%) originate from natural sources. Automobiles are the principal anthropogenic source.
- **Particulate matter** consists of small particles of solids or liquids. Dust from farming and construction sites, smoke, and soot are common types of particulate matter. More toxic materials may form particulates such as heavy aerosol forms of heavy metals, asbestos, and very fine particulates such as sulfates and nitrates (Figure 22.3). The health consequences of particulate pollution are substantial. An estimated 2% to 9% of the deaths in U.S. cities can be associated with particulate pollution; in addition, cities with high levels of particulate contamination have a 15% to 20% higher mortality rate than cities with less severe particulate pollution.
- **Hydrogen sulfide** (H_2S) is a natural product of some marshes and swamps, but it is also generated during petroleum processing, refining, and metal smelting. In high concentrations, H_2S can be toxic to plants and animals.
- **Hydrogen fluoride** (HF) is highly toxic and is a product of aluminum processing, coal gasification, and coal combustion.
- Other gases such as chlorine gas, formalin, or vapors from industrial plants can create air pollution problems.
- **Other hazardous gases** including chlorine and chemical used in agriculture and industry.
- **Lead** is now widely dispersed in the environment. Lead has been detected in samples of ice from Greenland's glaciers and fish tissue in remote Canadian lakes, far removed from lead sources. The main sources of lead contamination have been automobiles and leaded gasoline (now phased out in many developed nations). High concentrations of lead are found in soils along roadways and in urban dust. Birth defects and nervous system disorders have been attributed to lead poisoning; fetuses and children are especially susceptible.

Where and when are air pollution problems most severe?

- Urban and industrial areas typically have the most polluted air but areas far away from the sources of air pollution, such as the Arctic, have higher that expected concentrations of pollutants.
- Air pollution problems can be exacerbated by local topography and meteorological conditions. Urban and industrial areas located in valleys or adjacent to mountain ranges are more likely to experience **temperature inversions** that trap cool, polluted air below relatively warmer air, thus preventing the dispersion of pollutants (Figures 23.22 and 23.23).
- The potential for air pollution depends upon the four factors that are depicted in Figure 23.24:
    - the rate of pollutant emission per unit area,
    - the distance downwind that an air mass encounters pollutant sources,
    - wind speed,
    - height of mixing.

What is smog?

- The word ***smog*** (smoke-fog) was probably first used in the early 1900s to describe poor air quality, especially polluted air that reduces visibility in urban areas.
- There are two basic types of smog. **Photochemical smog** results from automobile emissions reacting with solar radiation to produce the brownish smog, characteristic of the smog of Los Angeles, California (Figure 22.7). **Sulfurous smog** is the gray smog generated in industrial areas such as London.
- Smog conditions change throughout the day as pollutants begin to accumulate and solar radiation intensifies (Figures 22.8 and 22.9).

What trends are expected in urban air pollution?

- Optimistically, more is known about the causes and prevention of air pollution so pollution problems should abate. This is the apparent trend in the more developed nations, especially if measured as air pollution generated per capita. However, there are an increasing number of individuals. Southern California has seen a reduction in emissions but still has the worst air pollution in the nation.
- Pessimistically, increasing population and increasing economic pressures will intensify the problem of air pollution, especially in developing nations. Developing nations are experiencing the most rapid increase in population and many of these individuals are moving to urban areas in the hope of economic improvement. Too often, the financial base to support pollution reduction programs is not sufficient. Mexico City is one of the prime examples of the unfortunate combination of insufficient pollution control, very high and increasing population, and climatic and topographic situations that exacerbate air pollution problems. In the US, many mid-sized cities are experiencing increased air pollution problems (Table 23..3)

What can be done to reduce air pollution in urban areas?

- Reduce automobile numbers and usage.

- Stricter automobile emissions standards and enforcement.
- Reduced or "non-polluting" cars (such as electric cars).
- Improved gasoline.
- Public transportation.
- Car pooling
- Increased controls and enforcement on industry and household pollution.

What air pollution legislation was passed in the Clean Air Act Amendments of 1990?

- The Clean Air Act Amendment of 1990 was a set of comprehensive federal regulations that addressed a suite of air pollution problems (acid rain, toxic emissions, ozone depletion, and automobile exhaust).
- To reduce acid rain, the Act mandated that the nationwide total of SO_2 emissions will have to be reduced by 50% to 10 million metric tons by the year 2000. To help accomplish this goal, incentive programs have been developed to encourage industries to use clean air technologies.
- A controversial incentive program is the "pollution credits" program that allows producers of large volumes of pollutants to purchase pollution credits from industries that produce fewer units of pollution.
- A 90% reduction of highly toxic emissions is a goal of the Act. Pollution abatement of toxic emissions depends upon the implementation of clean air technologies.
- Reducing emissions of stratospheric ozone destroying chemicals has been implemented in phased programs.
- Stricter regulations governing automobile exhaust and improving efficiency were incorporated into the Act.

What are air quality standards?

- Air quality standards (National Ambient Air Quality Standards) have been developed to limit the volumes of specific pollutants from industries, power plants, and automobiles.
- The effectiveness of relying on standards depends primarily upon developing standards that adequately reduce pollutant risks to human and environmental health and secondly upon the enforcement of those standards.
- Monitoring air quality standards provides a means of assessing the threat of air pollution to human health. Air quality monitoring of five pollutants is conducted (total suspended particulates, sulfur dioxide, carbon monoxide, ozone, and nitrogen dioxide) to obtain an Air Quality Index. Cities can use this index to issue health advisories (Table 23.4).

How much does air pollution prevention cost?

- The cost of pollution abatement is complex.
- There are costs paid by polluters to develop, install, and maintain pollution control technologies. With stricter regulations, the cost of pollution control increases; this cost is often transferred to consumers. The cost of pollution control increases incrementally; it is relatively inexpensive to reduce large volumes of pollution initially, but subsequent reductions in pollution are much more costly to implement.

- Concurrently, as pollution levels are reduced, the cost of pollution damage to human and environmental health is reduced. Less pollution reduces the cost of health care, food production, and the maintenance of environmental quality.

What is acid rain?

- There are **two types of acid rain**: "wet" which includes rain, snow, and fog, and "dry" deposition of acidic particulates.
- "Pure" rain water is slightly acidic (average pH of 5.6) (Figure 23.16).
- The **pH** of acid rain may range from below pH 5.6 to extremely acidic pH 1.5. The pH scale is logarithmic so changes in pH values represent changes in pH by orders of magnitude.
- Acid rain was initially perceived as a problem confined to the highly industrialized regions of northern Europe and the northeastern United States. Now the problem is widespread.
- Acid rain does not respect international borders. Sources of acid rain may originate in one nation and the effects may be experienced in an adjacent country. Examples are the United States and Canada or Poland and Slovakia.

What causes acid rain?

- Two types of emissions are responsible for most of the acid rain problems. Sulfur dioxide emissions (primarily from stationary sources such as power plants that combust coal) and emissions of nitrogen oxides (primarily from mobile sources such as automobile exhausts, as well as the combustion of other fossil fuels).
- In the atmosphere, these emissions combine with water or dust particles to form acidic precipitants.
- In the United States during the1970s the tonnage of SO_2 and NO_x emissions peaked at approximately 30 million metric tons per year; now emissions of these compounds is about 18 million metric tons per year (Figures 23.27 and 23.28).
- Eighty percent of the SO_2 and 65% of the NO_x emissions in the United States originate east of the Mississippi River.
- Industrial tall stacks were designed to disperse emissions and minimize local pollution problems, but by releasing emissions higher in the atmosphere, pollutants are carried further downwind and their residence time in the atmosphere is increased.

What are the environmental effects of acid rain?

- Acid rain can damage leaves and roots of plants and damage tissue of aquatic animals.
- The effects of acid rain depend, in part, upon the buffering capacity of the soil and water.
- Well buffered soils, such as those formed from carbonate rock, or lakes with carbonate beds, have a relatively high or neutral pH and are capable of counteracting some of the harmful effects of acid rain. However, pH may eventually be lowered with continual inputs of acid precipitants. Generally, soils of the arid portions of the western U.S. are well buffered (Figure 23.18).
- Poorly buffered soils and lakes, typically found in areas dominated by granitic rock, are most susceptible to acid rain problems. These conditions are most often associated with the humid, eastern U.S. Acidic conditions in soils cause nutrients to be easily leached and

certain metals to become toxic (e.g., Al, Pb, and Hg). In water, abnormally acidic conditions and aluminum toxicity are damaging to aquatic invertebrates, fish eggs, larvae, and gills and disrupt food chains by reducing nutrients available to photosynthetic algae.

- The addition of lime to lakes is a temporary, stop-gap measure to prevent or remediate acidification.
- Toxic heavy metals that have been become mobilized due to lake acidification can be passed up food chains and present a risk to persons that consume fish.
- Acid rain can dissolve statuary, monuments, marble and limestone buildings, and corrode metal. The economic and cultural losses due to acid rain are extensive.

How can acid rain and other air pollutants be controlled?

- The first priority should be reducing the volume of pollutants that is generated through conservation measures and improved efficiency of technologies that generate pollutants.
- Secondarily, pollutants should be collected, captured, or retained to prevent their dispersion into the environment, then the contaminants should be detoxified and the chemicals reused if possible.
- Particulates can be controlled by using settling tanks, controlling dust with covers and vegetation.
- SO_2 emissions from coal-fueled power plants can be reduced by using low-sulfur coal, processing coal to reduce the sulfur content prior to combustion, coal gasification, and the use of post-combustion scrubbers.
- Emissions from automobiles can be reduced by installing and maintaining pollution control devices, such as catalytic converters, improving fuel efficiency, and restricting old vehicles or vehicles that emit a large volume of pollutants.

Ecology In Your Backyard

- What air pollution problems exist in your region? What are the sources of the problems and how are the problems manifested in terms of human health and environmental quality? What pollutants are the most problematic?
- Many local or regional newspapers, television weather stations, and Internet resources can provide you with daily air quality information and health advisories.
- Via the Internet you can learn more about air pollution, or contact your state or local agencies that address air pollution issues. Search the EPA website (http://www.epa.gov/) key topic "air" for more information.

Links In The Library

- Hamill, P. 1993. Where the air was clear. *Audubon*, January/February 1993. pp. 38-49. An article about Mexico City's Air Pollution Problem.
- Hayes, R.D. 1993. Ravaged republics. *Discover Magazine*, March 1993. pp. 66-75. - An article about the Czech and Slovak Republics' air and water pollution problems.
- Malaspina, M., K. Schaefer, and R.Wiles. 1992. Air pollution solutions: What works. Report Number 1. *The Environment Exchange*, Washington, DC May 1992.

Ecotest

1. The source of most particulate, carbon monoxide, and hydrocarbon air pollutants is ____.
a. industrial tall stacks
b. uncontrolled emissions from power plants
c. natural sources
d. automobiles

2. What are secondary air pollutants?
a. nontoxic emissions that present minor health risks
b. any pollutant comprising less than 5% of the total pollutant load
c. pollutants that form from primary pollutants upon discharge to the atmosphere
d. pollutants that bioaccumulate and biomagnify in food chains

3. Which of the following gases is colorless, odorless, and has a high affinity for blood hemoglobin?
a. CH_4
b. CFC
c. NO_2
d. CO

4. The potential for air pollution problems to develop depends upon all of the following factors except:
a. the wind speed.
b. the distance downwind that an air mass encounters pollutant sources.
c. the rate of pollutant emission per unit area.
d. the buffering capacity of the regional climate.

5. The brown smog that is typical of Los Angeles, California is primarily _____ smog.

a. photochemical
b. sulfurous
c. polygenic
d. methanogenic

6. Currently, what is the approximate tonnage of both SO_2 and NO_x discharged to the atmosphere in the United States each year?
a. 18 million metric tons
b. 12 million metric tons
c. 5 million metric tons
d. 3 million metric tons

7. What region of the United States is the most susceptible to acid rain problems?
a. the arid region of southern California
b. the eastern region underlain by or soils derived from granitic rock
c. the mid-western region underlain by carbonate rock
d. the southwest near the Mexican border

8. A temperature inversion occurs when _______________; polluted air may accumulate during inversions.
a. continental air masses override maritime air masses
b. relatively cool air is trapped below relatively warm air
c. fossil fuel combustion causes as rise in ambient air temperature
d. relatively warm air displaces relatively cool air

9. An Air Quality Index is used to assess human health risks during air pollution episodes; this Index is based on pollutant levels for total suspended particulates, sulfur dioxide, carbon monoxide, ozone, and ____________________.
a. carbon dioxide
b. methane
c. chlorofluorocarbons
d. nitrogen dioxide

10. Which of the following statements about the cost of air pollution control is false?
a. The cost of pollution control decreases incrementally; it is expensive to reduce large volumes of pollution initially, but subsequent reductions in pollution are much cheaper to implement.
b. As pollution levels are reduced, the cost of pollution damage to human and environmental health is reduced.
c. With stricter regulations, the cost of pollution control increases; polluters often transfer this cost to consumers.
d. Less pollution reduces the cost of health care, food production, and the maintenance of environmental quality.

11. "Pure" rain water is:
a. slightly basic.
b. very acidic.
c. neutral.
d. slightly acidic.

12. Environmental effects of acid rain include:
a. damage to leaves and roots of trees exposed to acid rain.
b. nutrient leaching from poorly buffered soils.
c. corrosion of structures made of limestone or marble.
d. All of these are correct.

13. The word "smog" comes from:
a. the sound made by automobile exhausts, because automobile emissions constitute the major source of smog.
b. the first letters of sulfur, methane, oxygen, and gas, the principal components of smog
c. combining the words smoke and fog.

14. Most hydrocarbon air pollutants originate from:
a. natural sources.
b. automobiles.
c. industry.
d. mining.

15. What pollutant is a natural product of marshes and swamps?
a. sulfuric acid
b. sulfur dioxide
c. hydrogen sulfide
d. carbon monoxide
e. nitric acid

Chapter 24 - INDOOR AIR POLLUTION

The Big Picture

Indoor air pollution has been a health problem since humans first lived in enclosed areas with smoky fires. Today, many humans, especially those living in the more developed nations, spend the majority of their lives indoors. While indoors, we may be exposed to pollutants occurring in much higher concentrations than they occur outdoors. Only within the past three to four decades have homes and buildings been constructed in a manner that minimizes air exchange and continually recirculates the same air. While recirculating filtered air reduces energy losses associated with heating and cooling, in some situations human health can be at risk. Physical reactions to indoor air pollutants vary dependent upon the types of pollutants, exposure levels, and susceptibility of the individual. If a number of individuals are experiencing health problems and the common link between them is the building in which they reside or work, the building may have "sick building syndrome." Health risks from indoor air pollution range from minor irritations of the eyes, nose, or throat to chronic lung disorders, cancers, and death. A range of options is available to remediate indoor air pollution problems; simple bans on indoor smoking may suffice or complete remodeling of the structure may be required.

Frequently Asked Questions

What are the sources of indoor air pollution?

- Sources of indoor air pollution may be natural or due to human activities and man-made products.
- The most common indoor air pollutants include:
  - **bacteria and molds** (the warm, moist, dark, and grimy recesses of building ventilation systems may provide a favorable site for the growth of bacteria and molds)
  - **asbestos fibers** (asbestos has been used for decades as a flame-retardant insulation for walls and pipes)
  - **formaldehyde** (formaldehyde is used in particle board, plywood, and treated lumber as a wood preservative)
  - **tobacco smoke** (in enclosed areas, smokers place others at risk due to "secondhand" tobacco smoke)
  - **carbon monoxide** (a toxic by-product of combustion)
  - **radon gas** (a naturally occurring radioactive gas that is emitted from certain geologic formations)

Why is indoor air pollution a potentially greater health risk than outdoor air pollution?

- Several behavioral and technological factors combined cause indoor air pollution to be a potentially significant health risk to some individuals.
- Pollutant concentrations are often higher indoors than outdoors because indoor air is recirculated in an effort to maintain an optimum temperature and reduce energy losses from buildings.
- Air maintained at a moderate temperature and humidity may be conducive to microbial growth in airways and vents.
- Many humans, especially in the more developed nations, spend the majority of their lives in controlled climate buildings, thus increase their exposure to indoor air pollutants.

How are pollutants emitted to buildings?

- Pollutants may originate within buildings and accumulate due to poor ventilation. New upholstery, flooring, paint, fabrics, or even the structural material of the building often emit contaminant vapors.
- Pollutants may originate outside of the building and infiltrate via cracks in the foundation or vents and become transported throughout the building via the ventilation system. Because of air pressure differences inside and outside of buildings, contaminants may be drawn in and transported upward by a "chimney or stack effect."

Why are some people strongly affected by indoor air pollutants while others are affected mildly or not at all?

- Humans vary in their sensitivity to many pollutants.
- Varying sensitivity may depend upon genetic factors, life-style, age, type of pollutant, duration and concentration of exposure, and sensitization (repeated exposures).
- Children, older individuals, or individuals with respiratory, dermatological, or immunological ailments are especially susceptible.

What symptoms do people exposed to indoor air pollution exhibit?

- Irritation of the eyes, nose, throat, and skin.
- Dizziness, nausea, respiratory distress.
- If high concentrations of toxic pollutants are present, severe and acute reactions may occur, even death (for example, CO poisoning has recently attained national attention).
- Long-term exposure to indoor air pollutants may increase the risk of chronic diseases such as cancer and heart disease.

What is "sick building syndrome"?

- "Sick building syndrome" occurs when a number of people occupying the same building exhibit symptoms of exposure to indoor air pollutants or pollutants are detectable.

- "Sick building syndrome" may be intensified by exposure to multiple pollutants, other environmental stresses (such as poor lighting, noise, and unsuitable temperature and humidity), employment-related stress, overcrowding, or undetectable factors.
- "Sick building syndrome" not only affects human health but economic productivity, as well.

How do sick-building syndrome and building-related illness differ?
- Sick-building syndrome refers to conditions in buildings that cause illness but the causative agent is not identified.
- Building-related illness is a condition in which the causative agent is can be identified.

What is secondhand smoke?
- Environmental tobacco smoke, or more commonly, secondhand smoke is smoke that is either exhaled by smokers or emanates from burning cigarettes, cigars or pipes and affects non-smokers.
- The affected non-smokers or passive smokers suffer many of the same health complications as smokers.
- Secondhand smoke has resulted in bans of smoking in most governmental buildings and many stores and restaurants nationwide.

What is radon gas and why is it an indoor air pollution problem?
- Radon gas is a naturally occurring, radioactive decay product of radium 226.
- Radon is colorless, odorless, and tasteless; it can be detected with radiation detectors.
- Geologic formations that are high in uranium content tend to emit radon.
- Humans that live or work in houses or other buildings that are located atop geologic formations that contain radon are at risk of radon exposure.
- An estimated 7% of the homes in the United States have elevated radon levels.
- The Reading Prong region of Pennsylvania, New Jersey, and New York has one of the highest concentrations of radon exposure in the nation.
- Radon may enter homes and buildings in three ways:
  - as a vapor emitted from soil or rock into basements or lower floors and transported via chimney effect throughout the building
  - dissolved in groundwater and pumped into homes or buildings as water supply
  - in construction materials (such as building blocks or stone)
- The EPA estimates that approximately 10% of the annual lung cancer deaths (~14,000 people) in the United States may be attributable to radon (and its daughter products, such as polonium 218) exposure.
- The combined effects of smoking tobacco and radon exposure increase the health hazard by 10 to 20 times.
- In the United States, the average outdoor concentration for radon is 0.2 pCi/L (picocuries per liter); the average indoor concentration increases to 1.0 pCi/L. The EPA "action level" for radon is 4.0 pCi/L; at this concentration remedial action should be undertaken to reduce or eliminate radon exposure.

How can indoor air pollution be controlled?

- A variety of measures may be undertaken to control indoor air pollution. Control methods differ depending upon site conditions, pollutant types, building use, and economic considerations.
- Control may entail isolating and sealing off the pollutant source or removal of the source.
- Improved ventilation that dilutes pollutants and provides building inhabitants with a supply of clean air. Using counter-current heat exchangers for heating and cooling reduces energy losses while allowing for improved ventilation.
- Regular cleaning and maintenance of ventilation ducts could prevent such health problems as Legionnellosis (Legionnaires Disease) that results from exposure to bacteria that grow in grimy ductwork.
- In the case of secondhand tobacco smoke, restricting indoor smoking is behavioral control that is gaining acceptance nationwide.
- Installing smoke or CO detectors and conducting periodic inspections are simple and inexpensive ways to detect indoor air pollutant problems and to prevent hazardous situations from developing.
- By educating building occupants about the causes and symptoms of indoor air pollution and the options to remediate the problem and decrease exposure.

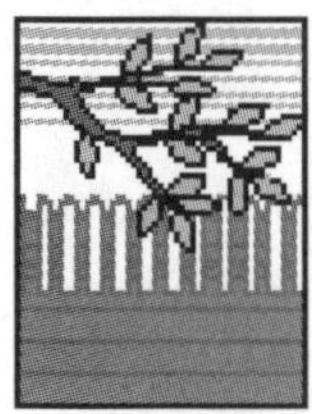

Ecology In Your Backyard

- Do any buildings on your campus or community suffer from sick building syndrome? Talk with campus building and housing personnel to see if problems have been detected and if so, find out if the problems have been rectified.

Links In The Library

- Nero, A.V., Jr. 1988. Controlling indoor air pollution. *Scientific American*. 258: 42-48.
- Page, S. 1994. Radon risks are real. *Garbage*. May, 1994. 29-31.
- Spengler, J. and K. Sexton. 1983. Indoor air pollution: a public health perspective, *Science* 221(4605): 9.
- Turiel, I. 1984. *Indoor Air Quality and Human Health*. Stanford University Press, California.

Ecotest

1. All of the following are common indoor air pollutants or pollutant sources except:
a. asbestos fibers.
b. bacteria and molds.
c. formaldehyde.
d. ethanol.

2. Air pressure differentials between indoors and outdoors may result in the upward movement of hazardous vapors into and throughout a building. This is called _____.
a. the chimney effect
b. an upwelling
c. an aerosol pathway
d. inflation

3. Exposure to radon is especially high in a region encompassing portions of Pennsylvania, New Jersey, and New York known as ________.
a. Dolly Sods
b. Pine Barrens
c. Nessmuk Plateau
d. Reading Prong

4. An estimated ___ percent of the homes in the U.S. have elevated radon levels.
a. 2
b. 7
c. 13
d. 23

5. By a factor of between 10 and 20, the combined effect of radon exposure and ______ increases the health risk that radon exposure alone would present.
a. tobacco smoking
b. asbestos
c. volatile organic compounds (VOCs)
d. chloroflourocarbons (CFCs)

6. The EPA estimates that ~10% or _______ of the lung cancer fatalities in the U.S. each year may be related to radon exposure.
a. 1, 200
b. 14, 000
c. 43, 000
d. 67, 000

7. Legionnellosis (Legionnaires Disease) was identified after an outbreak at an American Legion Convention at a Philadelphia hotel in 1976. Which statement best describes the outbreak?
a. The combination of cigarette smoke and exposure to high concentrations of carbon monoxide causes respiratory distress and chronic lung disease in convention attendees.
b. Hotel employees (kitchen staff and waiters) were exposed to spores of *Clostridium botulinium* from improperly preserved food and were stricken with botulism.
c. Convention attendees were exposed to bacteria that had colonized dirty air conditioning vents; several attendees died and many others became ill.
d. Exposure to high levels of radon gas resulted in the development of cancers of the respiratory tract in several convention attendees.

8. The effects of indoor air pollution and sick building syndrome often affect individuals differently because of all of the following reasons except:
a. differing sensitivity to contaminants.
b. differing preventative health care programs.
c. differing age, life-style, existing health status.
d. differing exposure concentrations and duration.

9. The EPA "action level" for radon is _____ pCi/L.
a. 4.0
b. 8.0
c. 12.0
d. 19.0

10. Acceptable methods used to control indoor air pollution include all of the following except:
a. isolating or removing pollutant sources.
b. eliminating counter-current heat exchangers used in most older buildings.
c. proper cleaning and maintenance of ventilation equipment and ductwork.
d. regulating behaviors that result in indoor air pollution (such as smoking).

11. Radon can enter buildings by all of the following except:
a. emission from rocks and soils into basements or lower floors.
b. being dissolved in groundwater and pumped into homes through water supply.
c. transpired by plants and trees surrounding a building.
d. introduced in construction materials.

12. What is passive smoking?
a. smoking less than one pack of cigarettes per day
b. the inhalation of ETS (Environmental Tobacco Smoke) by non-smokers
c. smoking in designated areas only
d. smoking but not inhaling

13. Indoor air pollution is potentially worse than outdoor pollution for all the following reasons except:

a. people spend more time indoors nowadays.
b. buildings tend to contain and recirculate air pollutants.
c. humans have evolved tolerances to most outdoor air pollutants.
d. indoor conditions provide habitats for organisms that are a source of air pollution.

14. Which statement about "sick building syndrome" is not true?
a. It can be intensified by crowded conditions within a building.
b. It can affect worker productivity.
c. The pollutant causing the sickness may be undetectable.
d. People can act as vectors, that is, they can transport the syndrome from one building to another.

15. Radon is:
a. colorless.
b. tasteless.
c. odorless.
d. none of the above.
e. all of the above.

Chapter 25 - OZONE DEPLETION

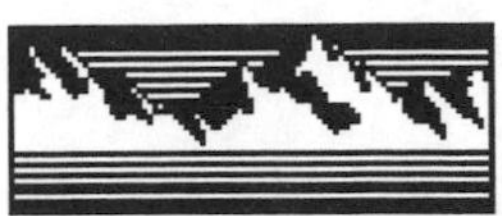

The Big Picture

Ozone is a pollutant in the lower part of the atmosphere (the troposphere), but in the upper atmosphere (the stratosphere), ozone forms a protective shield against the Sun's harmful ultraviolet (UV) radiation. Excessive exposure to UV radiation can damage the photosynthetic tissues of plants thus interfering with crop production as well as global productivity. In humans and other animals, excessive and long-term UV radiation exposure can damage eyes and skin. Since 1970, there has been an alarming decline in the concentration of ozone in the stratosphere during certain times of the year and at high latitudes. This decline has been attributed, in part, to pollutants such as a group of chemicals called chlorofluorocarbons, which disrupt the cycle of ozone formation and UV absorption. Ozone-depleting chemicals are also introduced into the atmosphere from volcanic eruptions. Ozone depletion is an international problem, and, as with many issues that must be addressed globally, there is a lack of consensus on how best to remediate the problem. The United States and several other developed nations have begun phasing out the use of many ozone-depleting chemicals.

Frequently Asked Questions

What is ozone?

- **Ozone (O_3)** in the troposphere is a pollutant, a product of photochemical reactions involving fossil fuels. **Stratospheric ozone**, however, forms the ozone shield which intercepts much of the harmful ultraviolet light **(UV)** energy entering the Earth's atmosphere.

Is ozone a form of oxygen?

- The principle form of oxygen in the atmosphere is the **biatomic molecule (O_2)**; **triatomic ozone (O_3)** occur in a much lower concentration.

What is ultraviolet light?

- There are **3 forms (wavelengths) of ultraviolet light** reaching the atmosphere (Figure 25.2); the effectiveness of the ozone shield in absorbing these wavelengths varies:
  - **UVA** (0.40 - 0.32 μm) is not absorbed by ozone, thus is unaffected by changes in the ozone shield.

- **UVB** (0.32 - 0.28 μm) is partially absorbed, thus depletion of the ozone shield permits more UVB to reach the Earth's surface.
- **UVC** (0.28 - 0.10 μm) is strongly absorbed by ozone, thus as long as a minimum of ozone is present in the atmosphere, UVC will be absorbed. The intensity of UVC reaching the Earth's surface will not be affected by reductions in the ozone shield.

What does ultraviolet light do when it passes through the ozone layer?

- Ultraviolet light dissociates biatomic oxygen (O_2) into two **monoatomic oxygen atoms (O)**.
- The monoatomic atoms combine with other diatomic oxygen molecules to form triatomic ozone (O_3).
- Additional inputs of UV cause O_3 to split into O_2 and O, which may recombine to form O_3; UV radiation is absorbed in this process.
- If ozone-depleting chemicals are present (such as CFCs or other molecules containing chlorine), a catalytic chain reaction occurs in which ozone will not be reformed.

How can we tell how much ozone is present in the stratosphere?

- Ozone is measured in Dobson Units (DU). Based on measurements taken between 1957-1970 in the Antarctic, we may assume that baseline stratospheric ozone concentrations are approximately 300 DU.

How do we know that ozone is declining in the stratosphere?

- Since 1970, there has been a trend toward decreasing ozone concentrations (Figure 25.7). This is especially evident over the Antarctic continent. Concentrations since 1970 have ranged from 250 DU to 90 DU, indicating a thinning of the ozone shield especially during the Antarctic spring.

What causes the decline in ozone in the stratosphere?

- The decline of the ozone shield has been attributed to the emissions of chlorofluorocarbons (CFCs) and similar air pollutants, and natural sources of chlorine in the atmosphere (i.e., from salt in the ocean).
- UV light causes CFCs to dissociate into free chlorine (:Cl) radicals, which combine with disassociated ozone to form ClO and oxygen (O_2), thus depleting the stratosphere of ozone
- CFCs are being used as refrigerants in freezers, refrigerators, and air conditioning systems, as industrial cleaning agents, and as a blowing agent for some kinds of foam and aerosol cans.

Where is the ozone depletion worst?

- The depletion of stratospheric ozone is especially problematic over the Antarctic.
- During the Antarctic winter, the Antarctic air mass becomes isolated from the rest of the atmosphere.
- As the Antarctic spring approaches, within this vortex of rotating air, polar stratospheric clouds develop.

- As the clouds form, nitrogen oxides (NO_2) condense with frozen sulfuric acid (H_2SO_4) or water vapor and precipitate out of the stratosphere.
- With the loss of nitrogen oxides, an important reaction (formation of chlorine nitrate or $ClO{\cdot}NO_2$) can no longer occur.
- Chlorine free radicals (:Cl), which are created from CFCs by UV light, bind rapidly with ozone to form chlorine oxide (ClO) and oxygen.
- This mass of ozone deficient air over Antarctica ("the Ozone Hole") may migrate toward the equator and present risks to the biota of the affected region.
- In the tropics and middle latitudes, stratospheric ozone depletion has also been severe (10 %) and is correlated with volcanic eruptions such as the Mount Pinatubo in the Philippines and El Chichon in Mexico, which inject sulfur dioxide into the stratosphere. This natural source of ozone depleting chemicals is currently being investigated.

Does the loss of ozone harm people?

- Ozone depletion and increased UV exposure poses risks for environmental and human health.
- Direct effects of increased UV exposure on human health include: increased incidence of skin cancers, eye disease (cataracts), and weakened immune systems.
- However, it may be difficult to attribute a specific UV-related illness to any one cause since risks vary according to an individual exposure, sensitivity, and longevity.

What precautions can individuals take to reduce the health risks associated with UV exposure?

- Minimize exposure to mid-day sun; keep to the shade
- Use sunblock creams of SPF 15 or greater
- Wear UV-protective sunglasses
- Avoid using tanning salons and tanning lamps
- Learn about your individual risks based on skin type and family history of skin cancer; consult a physician and/or internet websites focusing on UV protection.

Does the loss of ozone harm the Earth's ecosystems?

- Increased UV interferes with photosynthesis. Oceanic food chains may be altered if primary productivity by phytoplankton and photosynthetic bacteria is diminished.
- Loss of oceanic productivity may result in an increase in atmospheric CO_2 concentrations, thus exacerbating the "greenhouse" effect and the problem of global warming.
- Damage to crops may reduce yields. Reductions in food production could have major social, political, economic, and human health consequences.

What is being done about the ozone depletion in the stratosphere?

- Fortunately, releases of new CFCs to the atmosphere have been reduced since 1992 although the CFCs already present in the atmosphere will continue to deplete ozone.
- The **Montreal Protocol (1987)** was an agreement among 145 nations to phase out emissions of CFCs. Not all nations that manufacture or use CFCs participated in the

agreement (for example, China and India, two countries with growing human populations).

- In developing nations, the demand for refrigerants is expected to increase. Unless safe and economical substitutes for CFCs become available, it is unlikely that these nations will voluntarily reduce CFC emissions.
- Controversy exists over the cause of ozone depletion. Proponents of the hypothesis that stratospheric ozone depletion is anthropogenic cite evidence for CFCs in the stratosphere and changes in the concentrations of hydrogen chloride and hydrogen fluoride in excess of amounts expected from volcanic activity.
- Critics propose that evidence for CFCs in the stratosphere does not exist and that increased concentrations of chlorine in the atmosphere may be due to recent volcanic activity or chlorine aerosols from the ocean. Such controversies arise as a result of apparently poor communication among scientists, industry, government, and the general public. Further, understanding global environmental problems is inherently complex and difficult.
- Substitutes for CFCs being suggested are hydrochlorofluorocarbons (HCFCs) and hydrofluorocarbons (HFC), which have a shorter half-life in the stratosphere than CFCs, and thus will deplete less ozone. However, both chemicals are currently expensive to produce and can't be used in current refrigeration systems. It is estimated that it will cost $385 billion for updating the older equipment to handle HCFCs and HFCs.

Ecology In Your Backyard

- What is your contribution to the CFC problem?
- What appliances may still contain CFCs in your house, apartment, or dormitory?
- As consumers, we may be indirectly contributing to the production of CFCs through our purchases of frozen foods (especially if they have been produced in nations that have not accepted the Montreal Protocol) and our utilization of manufactured goods that use or contain CFCs (such as car air conditioner systems). Do you think that it is realistic to require that products that we purchase should be "certified" CFC-free?
- In your community, where would you go to discard an old refrigerator or air conditioner that may contain CFCs?
- Does your local landfill accept these items?
- Who is responsible for assuring that CFCs are removed from coolant coils?
- Do you consider the health risks of UV exposure when engaging in outdoor recreational activities? What precaution do you take? Consult websites regarding UV exposure such as: http://www.fda.gov/cdrh/tanning.html
- Check with local meteorologists to determine if daily UV exposure risks are published on a local website.

Links In The Library

- de Gruijl, Frank R. 1995. Impacts of a projected depletion of the ozone layer. Consequences:, Vol. 1, No. 2, pages 12-21.
- Tevini, M.(ed.) 1993. *UV-B Radiation and Ozone Depletion*. Lewis Publishers, Boca Raton, Florida.
- van der Leun, X.Tang, and M. Tevini. 1994. Environmental effects of ozone depletion: 1994 assessment. United Nations Environment Programme (UNEP), Nairobi, Kenya (available from UNEP Ozone-Secretariat, P.O. Box 30552, Nairobi, Kenya.)

Ecotest

1. The depletion of stratospheric ozone results in an increase of which ultraviolet wavelength subdivision?
a. UVA
b. UVB
c. UVC
d. UVD

2. Ozone concentrations are measured in Dobson Units (DU). DU values have declined since 1970. Between 1957 and 1970, DU values measured at the British Antarctic Survey's Halley Bay site averaged _______ DU.
a. 500
b. 400
c. 300
d. 200

3. Which statement best describes ozone formation in the stratosphere?
a. Ozone is created when infrared light strikes oxygen molecules in the stratosphere.
b. Ozone is formed when fossil fuels are burned, releasing organic molecules that are broken down by UV radiation.
c. Ozone is formed when UV radiation strikes a CFC molecule, releasing a chlorine atom that steals an oxygen atom from chlorine nitrate $ClONO_2$.
d. Ozone is formed when UV light strikes a diatomic atom of oxygen and forms two monoatomic atoms of oxygen, which then combine with other diatomic molecules of oxygen.

4. Which statement about polar stratospheric clouds is incorrect?
a. Polar atmospheric clouds form within the Antarctic polar vortex.
b. During the Antarctic winter, cloud particles within polar stratospheric clouds serve as chlorine sinks.
c. Polar stratospheric clouds drift toward the middle latitudes during the Antarctic summer, thus depleting the Antarctic of its protective ozone shield.
d. Increased UV radiation during the Antarctic spring liberates free chlorine from cloud particles, thus initiating ozone-depleting catalytic chain reactions.

5. The possible consequences of ozone depletion include all of the following except:
a. an increase in the eutrophication of natural waters.
b. an increase in the incidence of skin cancers and cataracts and a weakening of immune systems in humans.
c. a reduction of marine phytoplankton abundance.
d. a decrease in food crop production.

6. What is one of the main obstacles to using HFCs, HCFCs, and helium as substitutes to CFC?
a. There is disagreement among technical advisors to the United Nations/World Health Organization.
b. These substitutes have been demonstrated to be ineffective in controlled scientific tests.
c. These substitutes are too expensive to produce and are not economically feasible at present.
d. Chemical companies prefer to develop propane as the preferred CFC substitute.

7. Which nations have not chosen to participate in the Montreal Protocols, an international agreement to reduce global emissions of CFCs?
a. Canada
b. United States
c. European Economic Community
d. China and India

8. What can individuals do to reduce their exposure to UV radiation?
a. minimize mid-day exposure to the sun
b. use sunblock of SPF 15 or greater
c. wear UV protective sunglasses
d. all of the above

9. If the emission of ozone-depleting chemicals were to cease completely today, what would be the probable consequence?
a. Stratospheric ozone concentrations would rapidly increase
b. Stratospheric ozone concentrations would be maintained at current levels
c. Stratospheric ozone concentrations would continue to decline
d. Stratospheric ozone levels would not be detectable

10. Which is a true statement about ozone depletion in tropics and middle latitudes?
a. Stratospheric ozone depletion has not been detected in the tropics and middle latitudes.
b. Stratospheric ozone depletion in the tropics and middle latitudes seems to coincide with volcanic activity.
c. Stratospheric ozone depletion is greatest in the tropics and middle latitudes, but this region is buffered from UV radiation by dense atmospheric gases.
d. Stratospheric ozone depletion has not been verified in the tropics, middle latitudes, or polar regions.

11. What is the molecular formula for ozone?
a. $3O_2$
b. O_2H_2
c. O_3

12. A natural source of ozone-depleting chemicals might be:
a. changes in ocean temperature.
b. volcanic eruptions resulting in sulfur dioxide.
c. methane production as a result of wastewater treatment.
d. There are no natural sources of chemicals that deplete ozone.

13. Nitrogen oxides are:
a. beneficial to the ozone layer, because they removes Cl^- free radicals.
b. detrimental to the ozone layer, because they causes ozone to dissociate into two smaller oxygen molecules.
c. not a problem because they have little to no effect on ozone levels in the atmosphere.
d. detrimental to the ozone layer, because they cause acid rain.

14. Ozone depletion is worst over:
a. the Arctic.
b. Hawaii.
c. Western Europe.
d. Antarctica.

15. Ozone can be a pollutant in the atmosphere, but only within which part of the atmosphere?
a. stratosphere
b. ionosphere
c. corona
d. troposphere

Chapter 26 - ENVIRONMENTAL ECONOMICS

The Big Picture

In this chapter, the economic values of environmental resources are considered. Environmental economists attempt to understand how much people value natural resources, especially commonly held resources such as forests, fish, water and air. In some cases, a strict dollar amount can be estimated for a resource (i.e., the value of all the timber in a forest), but in other cases the value of a resource is difficult to estimate (i.e., the aesthetic value of that same forest). People don't always agree on the value of a resource, because they use a different method to assess the future value of the resource, which economists refer to as the discount rate (the ratio of future value to current value). The value of a forest in the future may be low to some people, like timber industry employees, who want to harvest all the trees in a forest and invest the profits in another business that gives a higher return on investment. The future value of that forest may be much greater for another person who enjoys viewing the scenic vistas, hunting or fishing in that forested area. It is possible that some may believe that the future value exceeds the current value of the forest. These principles can be applied to pollution control planning, and the costs and benefits of different levels of pollution control can be estimated. Using economic approaches to examine the tradeoffs implicit in environmental decision making can help clarify the values held by different people for the future value of resources, but they may not completely resolve the differences among the people.

Frequently Asked Questions

What is economics?

- Economics is a science in which the costs and benefits of goods and services are estimated and evaluated.

What is environmental economics?

- Environmental economics is a science that attempts to analyze the value of ecological goods and services (animals, plants and the roles they play in an ecosystem).
- For example, it is important to understand the factors that affect the cost of pollution control and to evaluate the aesthetic values of wildlife and fisheries.

What do economists mean by utility?

- Economic utility is the same as the value or worth of a resource.

What is meant by the discount factor?

- The discount factor is the ratio of future utility (or worth) to present utility (or worth).
- The discount rate can be estimated very differently by different people:
  - Conservationists hold that the future value of a resource is equal to (or even greater than) the current value (i.e., the discount factor is 1.0 or greater).

Conservationists often state that a natural resource is "priceless" and should be protected for future generations, thus expressing their view that the discount rate is > 1.0.

- Capitalists view the future value of a resource as less than the current value (discount rate is less than 1.0). Capitalists attempt to maximize profits now; they care little for the future value of a resource, because they may not be alive to enjoy the benefits.

What is meant by the term "externalities"?

- External costs are referred to as externalities. These are the costs not paid directly by the manufacturer of a good or provider of a service. They are real costs, however, borne by someone other than the manufacturer.
- Internal costs are those that are paid for directly by the maker of a product.
- For example, the cost of purchasing the raw materials for the product are internal or direct costs. These costs are reflected in the selling price of a good or service.
- Pollution is a cost that is paid for by society as a whole and not the producer of a product, so it is referred to as an externality or indirect cost.
- Pollution control involves internalizing externalities, something that individual producers are unwilling to do without punishment (regulation and fines) or rewards (tax incentives and pollution trading rights).

What is the tragedy of the commons?

- The tragedy of the commons occurs when a commonly owned resource is harvested to a level that is no longer profitable to exploit.
- Commonly held resources, such as fish populations in the sea, are not privately owned, but rather owned by an entire country or a group of people. On a global scale, the atmosphere could be considered a commons.
- Many resources fall into this category including fishery, forestry, air, water, and genetic resources.
- Garret Hardin popularized this idea in 1968 with his *Science* magazine article, "The Tragedy of the Commons", one of the most widely read and cited papers in all of science.
- Hardin described a hypothetical case in which an individual's self-interest in grazing his cattle on a common pasture led to a situation in which all individuals in the common-pasture arrangement lost money. This individual cared little about the external costs of grazing the cattle at too heavy a rate.
- This can best be summarized as: "What is profitable for me in the short-term will eventually lead to an unprofitable outcome for everyone in the long-term."
- According to Hardin, without some kind of government action to regulate the behavior of individuals, a tragedy for the common resource cannot be averted.

How does the low growth rate of natural resources cause overexploitation?

- The low population growth rate of certain natural resources relative to the economic rate of growth makes them susceptible to over-exploitation.
- This is because rates of growth of populations rarely exceed about 5 % per year (see chapter 5), whereas investments in the stock market or a bank can grow at 12 % or greater each year.

- This means that the decision to harvest a growing population at a sustainable rate will have a lower return on investment than a policy of harvesting at a maximum rate over a short period of time.
- If the population is harvested at the maximum rate, then it is possible to maximize short-term profits, sell the harvesting equipment, and get out of the business of harvesting the resource, although this will lead to a long-term decline in the population.
- Many companies that harvest natural resources (forestry and fisheries) have chosen the second profit-maximization approach at the expense of natural resource sustainability.
- Old-growth timber harvests on private land in the Pacific Northwest (Headwaters Area Old Growth Forest in Humboldt County, CA) are being exploited in just this way by Pacific Lumber Company, with non-sustainable clear-cutting methods being used to remove as much of the forest as possible quickly. The "forest liquidation" profits are being used by the Maxxam Corp (owner of Pacific Lumber Company) to pay off debts in the junk bond market. After all the old growth has been cut down, it would not be profitable to grow more trees (it takes 250 years to grow back an "old growth" tree, and second growth trees do not yield a harvest that is equally profitable). The logging companies are ready to pull out of the Pacific Northwest when the trees are all cut down.

Why have whales been exploited to extinction?

- The exploitation of the great whales is an example of "The Tragedy of the Commons".
- The whales are a commonly held resource, and the whalers have no assurance of a future whale harvest.
- Thus, the value of whales to the whaling company is less in the future than it is now (discount factor is less than 1.0).
- Even if they are allowed to reproduce and are harvested sustainably, whaling is a bad investment from a strict economic viewpoint, because the whales will return at most a dividend of only 5%.
- If the whales are harvested to extinction, and the company's ships are sold off, the profits can be invested in the stock market, where they will earn 12 % (or more).
- From a strict economic viewpoint, it is a no-brainer: harvest every last whale and get out of the whaling business. This is largely how the whaling industry has progressed and in the progress we have lost some whales to extinction.
- Certainly, many people disagree with this strict economic viewpoint. For instance, Greenpeace supporters have a different view of the discount factor (greater than 1.0) and the external costs associated with whaling than a whaling company.

How do people value the future of natural resources?

- The future value of natural resources can be estimated very differently by different people:
    - Conservationists often hold that the future value of a resource is equal to (or even greater than) the current value. Conservationists often state that a

natural resource is "priceless" and should be protected for future generations.
    - Others may hold the view that the future value of a resource is less than the current value. Such people may attempt to maximize profits now; and care little for the future value of a resource, because they may not be alive to enjoy the benefits.
- These two viewpoints represent extremes in the opinions held by people on the subject of natural resource conservation. In both cases, the opinions are based on the future value of a resource, which is not known with certainty.
- The value to future generations of people is uncertain because they may not want to have the resources that are being conserved, or they may value the resources even more than we do (for example, wilderness areas may become more valuable in the future when there is less of it than there is now).
- It is clear that if a resource cannot be replaced, such as in the case of genetic diversity, then the future value should be assumed to be high.

What are the costs of pollution?

- The cost of pollution control is high in real dollars: In the U.S. it is currently about $115 billion/year, or about 2% of the GNP.
- The cost of pollution control in the U.S. is estimated to be $30-$60/year for a median income family, a relatively small amount per family.
- This cost does not include the cost of protecting endangered species, wetlands, forests, and declining fish and wildlife populations.
- Compare these amounts with the budget of the U.S. Environmental Protection Agency, which is $ 6 billion, far less than is required for pollution control alone.

What is meant by marginal cost?

- The cost of producing an additional unit of something when you are at the break even point, whether it is a widget or a unit of pollution abatement, is the marginal cost.
- With goods like widgets (a fictional good recognized only by economists), as more units are produced, the marginal costs decline, because the initial fixed costs to set up equipment are already paid and the resources needed to make the widgets are unlimited.
- With pollution control, there is an increase in marginal cost with additional units of pollution removed. This is because the recovery cost increases as the pollution "resource" declines.
- For example, it costs $0.05/kg to remove pollution when 20 % of the pollutants have been removed. However, it costs $0.49/kg to remove the pollutants when 80 % have been removed.
- The marginal cost of removing 100 % of the pollutants is very high, perhaps infinite, by extrapolation of the above relationship.

What are the economic benefits of aesthetic values?

- Cultural and aesthetic values (such as the sight of a bald eagle) are difficult to place a dollar value on, so they are referred to as **intangibles**.

- One of the goals of environmental economics is to develop a method of aesthetic evaluation that is quantitative and credible.

What is risk-benefit analysis?

- The comparison of the risk associated with an activity and the benefits of that activity is called risk-benefit analysis.
- Risk is the probability that some unwanted event will occur (death or injury). The probability ranges from 0, being no risk, to 1.0, being certain risk.
- Riskiness of future events influences the value we place on things now, and we are often forced to weigh the future risk against current benefits of some action.
- Benefits are outcomes that are judged to be favorable.

How are risks estimated?

- Risks are estimated by counting the number of unfavorable outcomes and dividing this by the total number of outcomes measured.
- For example, if 300 people died in plane crashes in a year with 100,000,000 passengers on all flights , the risk of death is said to be 3 in 1,000,000.

What is greater: the risk of dying while flying across country or dying in a motor vehicle accident when driving across country?

- The risk of dying in a car crash is far greater (2.2 in 10,000) than flying in a plane across country (3 in 1,000,000). Far more deaths occur each year from car crashes than plane crashes, when compared on a per-passenger mile basis.
- Many people are afraid to fly because they fear the risk of a crash, but these same people will drive because they feel more in control in a car. These people are actually putting themselves at greater risk by not flying.
- Risk is not perceived in the same way by people, and this perceptual difference accounts for their behavior. Some people fear flying because it is a novel form of transportation and some people do not want to give up control to another person.

How can we use risk analysis to make decisions about the acceptable level of pollutants?

- It is unlikely that the risk of injury and death from pollution will ever be reduced to zero, because it is too costly (Figure 26.4), but we may agree on an acceptable level of pollution.
- Total pollution costs (TC) are the sum of the loss from pollution damages (LPD, an external cost) and the costs to control pollution (CCP, an internal cost), i.e. TC = LPD + CCP.
- Initially, the total costs of pollution are high, mainly because of high cost of pollution damage.
- As pollution control measures are instituted at the request of governments acting on behalf of the people who suffer the pollution damages, the costs due to the loss of pollution damages fall, but the cost to control pollution rises.
- There is a minimum level of TC, where the shrinking LPD and the growing costs of CCP are approximately equal (where the lines representing these costs cross). Some argue that we should control pollution costs only to this point, to minimize overall costs.

- Others argue that we should reduce pollution to zero, which may be technically impossible and extremely costly (although LPD costs would be zero, the cost to control pollution (CCP) would be excessive).
- There has been much disagreement on what level of pollution control should be required and what is an "acceptable" risk of pollution.
- The level of pollution risk that is acceptable is more a matter of human values, morals, and ethics than a matter of science, because there will always be some risk at any level of pollution above zero, so someone will always be injured or die if any pollution occurs.

What level of risk from pollution is "unreasonable"?

- Different people will find different levels of risk acceptable:
- For example, consider a plastic plant making chemicals and dumping the waste into a coastal bay.
    - An industrialist who owns this plant will not consider pollution near the plant a great risk to himself, especially if he lives far away from the plant in a clean environment. The level of pollution that is acceptable to him will be relatively high.
    - In contrast, a woman living near the plant will have a different idea of what constitutes "acceptable" risk; She will demand a far lower risk and she may even insist on a risk of zero, which is impossible because the cost of achieving it is infinite, according to economic theory.
    - The industrialist will bear the total cost of pollution control, and so he will take the minimum steps to abate pollution, unless required by law.
- The term "unreasonable level of risk from pollution" is used in the Toxic Substances Control Act to define a level of risk which is prohibited by law.
- In practice, it is difficult to define what level of risk is "unreasonable"; it is a question of one's values.
- An acceptable level of pollution depends entirely on the value humans attach to human life.

What is the value of a human life?

- How much is a human life worth? This is the economic question we must address, even though everyone would agree on ethical grounds that all human life is priceless.
- Even though we don't think that we place a value on a human life, we routinely make cost/benefit analyses of life's daily risks and thus we do place a value on our own human life.
- We often must make an economic assessment rather than an ethical assessment of the value of human life.
- Using the rationale of an economist, one simply must ask people what they would pay to have their life extended.
- By one account, it is $32,000 per life saved, according to a study by the Rand Corporation. Researchers asked people what amount they would be willing to pay to set up a mobile heart attack screening system, with increased numbers of ambulances and improved intensive care units for heart attack victims. Such a

system would help them extend their lives and anyone else's who had a heart attack (the leading cause of death in the US).

What is a policy instrument?

- A policy instrument is a method of altering the behavior of individual people or organizations so that they will conform to the wishes of society at large.
- Policy instruments can include moral suasion (publicity and political persuasion), direct control (regulations), and market processes (taxes, subsidies, licenses, deposits, and market trading of pollution rights).
- Examples of policy instruments include regulations that limit fishing through harvest quotas or licensing programs; regulations limiting pollution amounts, laws requiring pollution discharge permits with fines for excessive pollution, and pollution trading rights sold on commodities exchanges.
- Which policy instrument is chosen depends greatly on the political opinions in the society.
- In the U.S., direct controls are often used to establish regulations for the maximum amount of pollution or the harvest quota, with fines for violating the limits.
- In Europe, it is much more common to have taxes levied according to the amount of pollution generated by an individual or industry so that there is an economic incentive to control pollution.
- The most effective type of policy instrument is a license because it causes people to act as if they have a stake in the future of the resource (Table 26.4)

Ecology In Your Backyard

What is the value you place on your own life? Find out what the risk of death or injury is from each of the following activities that you may do:

- Do you ride a bike?
- Do you skateboard?
- Do you drive a car?
- Do you ride a motorcycle?
- Do you bungee jump?
- Do you drink alcohol to excess?
- Do you do some other risky activity (explain)?

- All of these put you more at risk than does pollution, yet most people show more concern about pollution than they do for these activities. Part of the explanation for this difference is that when you drive or do one of these activities above, you feel that you are in control (even though this is a false sense of security - you may still die while driving, if you are hit by a careless driver). Pollution comes from an outside agent (an industry) that may be far away. Over this one has little control. Consider the risk of death or injury you would experience if you were to go snow skiing for the first time. Now consider the emotions that you would feel if you were kidnapped at gunpoint and forced to go skiing when you didn't want to. The actual risk of injury and death is the same in both cases, but you are angrier about the second risk. (See Rodericks, 1992)

- What role does lack of control over the source of pollution play in your assessment of the risk of a pollutant in your local area?

Links In The Library

- Allman, W. 1985. Staying alive in the Twentieth Century. *Science 85* 5(6): 31. October 1985.
- Ames, B., R. Magaw, and L. Gold. 1987. Ranking possible carcinogenic hazards. *Science* 236(4700): 271.
- Costanza, R. 1991. *Ecological Economics; The Science and Management of Sustainability*. New York: Columbia University Press. 525 pp.
- Hardin, Garrett. 1968. The tragedy of the commons. *Science* 162: 1243 - 1248.
- Jansson, A., M. Hammer, C. Folke, and R. Costanza. 1994. *Investing in Natural Capital; The Ecological Approach to Sustainability*. Washington D.C.: Island Press. 504 pp.
- Rodericks, Joseph V. 1992. *Calculated Risks: understanding the toxicity and human health risks of chemicals in our environment*. Cambridge University Press, Cambridge, England, 256 pp.
- Schmookler, A.B. 1993. *The Illusion of Choice: How the Market Economy Shapes Our Destiny*. State University of New York Press, Albany 349 pp.

Slovic, P. 1987. The perception of risk. *Science* 236(4799): 280.

Ecotest

1. In which of the following businesses should you invest your money: 1) shrimp harvesting (shrimp that reaches maturity in 1 year with an annual growth rate of 10 % and brings about $4 per pound) or, 2) red drum harvesting (a red drum takes 5 years to mature, has a population growth rate of 2 %, and brings about $1.50 per pound)?
a. red drum
b. shrimp
c. Both would be equally profitable.

2. If the cost of the removal of PCBs from a factory effluent is $10.00/pound when there is 10 % of the pollution removed, what is likely to be the cost when 50 % has been removed?
a. $5.00/pound
b. $15.00/pound
c. $1.00/pound
d. None of these are correct.

3. The harvest rates for a population of marine fishes continues to increase when there are no restrictions on harvest. What effect would issuing a limited number of commercial licenses have on the harvest rates?
a. Harvest rates will continue to go up.
b. Harvest rates will go down.
c. Harvest rates remain the same.

4. When is the total cost of pollution (TCP) the smallest?
a. when no pollution controls are in place
b. when pollution is reduced to zero
c. at an intermediate level of pollution control where the cost of pollution control is approximately equal to the cost of pollution damage

5. Which of the following is an "externality" in the harvest of mahogany trees described in the book?
a. the cost of replanting new mahogany trees to replace those harvested
b. the construction of new roads to bring the trees to market
c. the change of the use of land where trees formerly were found to agriculture
d. the alteration of forests so that open areas where new trees could sprout are now rare
e. All of these are external costs.

6. How is the future valued by conservationists who want to stop the logging in the Headwaters Forests?
a. The future is less valuable than the current value of the trees.
b. The future value is more than or equal to the current value of the trees.
c. The future value of the trees is irrelevant to the protection of the trees.

7. In which of the following activities is the risk of dying the greatest for the average person?
a. Flying across the country
b. Driving across the country
c. Breathing polluted air
d. Playing football
e. All of these activities expose the average person to the same risk of dying.

8. How much does pollution control currently cost per family per year in the U.S.?
a. $ 30-60
b. $ 120
c. $ 1000
d. $ 0.50

9. Which of the following is an example of an environmental intangible?
a. the value of the lumber in the Headwaters Forest
b. the scenic view of the Headwaters Forest
c. the value of the commercial salmon harvest that will be lost if the Headwaters Forest trees are cut down, due to river siltation
d. All of these are intangibles.

10. Which type of policy instrument is usually the most successful in controlling pollution?
a. direct control
b. moral suasion
c. permits and licenses to pollute
d. pollution control subsidies

11. Pollution control costs currently make up what percent of the Gross National Product of the United States?
a. 2 %
b. <1 %
c. 5 %
d. 10 %

12. If the loss from pollution damages is 10 million dollars, and the cost of pollution control is 20 million dollars, what would the total pollution costs be?
a. 30 million dollars
b. 10 million dollars
c. 20 million dollars
d. none of these

13. With pollution control, marginal costs:
a. increase as additional units of pollution are removed.
b. decrease as additional units of pollution are removed.
c. increase until a point when pollution control mechanisms are in place, then decrease.

14. When the discount factor of a resource is greater than one,
a. the future worth of the resource is greater than the present worth.
b. the present worth of the resource is greater than the future worth.

c. the future worth of the resource is less than the present worth.
d. the resource will depreciate in worth over time.

15. Capitalists optimistically view the discount rate of a resource as
a. greater than 1.
b. equal to zero.
c. less than 1, but greater than zero.
d. less than zero.

Chapter 27 - URBAN ENVIRONMENTS

The Big Picture

For centuries, cities have depended on their rural surroundings as support systems, providing food, water, building materials, and other essential goods and services. For centuries, this dependence was tangible. Modern cities, however, have been erroneously perceived as independent and self-sufficient. Cities function like ecosystems, drawing upon resources from adjacent ecosystems and exporting resources and wastes. If the environmental and human health qualities of cities are to be enhanced and maintained in a sustainable manner, city planners and citizens will have to recognize the urban environment as an ecosystem and develop a better understanding of urban ecology. In many of the most vibrant and viable cities of the world, there is a long history of designing urban landscapes with nature. In other cases, natural landscapes in cities have been sacrificed for short-term economic gain or for short-term solutions to socioeconomic and overpopulation problems; this is especially evident in some cities of developing nations. By the year 2000 one-half of the world's population will be living in urban areas; it will be essential to incorporate basic ecological concepts and designing with nature into urban planning.

Frequently Asked Questions

How has the role of cities in human history undergone change?

- Cities have long been considered the antithesis of the natural environment and environmentalism. This stems from the perception that in urban settings, the separation between humans and "Nature" seem to be greatest.
- However, more humans live in urban areas than ever before. In the United States, 70% of the population lives on 3% of the land area and 75% live in urban areas.
- Globally, 45% of the human population lives in urban areas.
- Cities are growing larger. In 1995 there were 23 metropolitan areas with a population of 8 million or greater; by 2015, 36 such cities are expected.
- Given these changing demographics, the ecological relationship of cities with their rural surroundings will have to be carefully considered and planned for in order to preserve or enhance environmental quality and quality of life within urban areas and in their rural support systems.
- Understanding urban ecology requires an understanding of the ecological needs of cities and the effects of cities on the surrounding countryside.

In what ways is a city an ecological system?

- The ecosystem concept can be extrapolated to cities. Ecosystems are comprised of a biotic community and the abiotic factors that drive or affect the system. In natural ecosystems, energy is derived primarily via photosynthesis and the exchange media are nutrients. In cities, energy is derived primarily from fossil fuels and the exchange media are goods, services, and waste products.
- Ecosystems are not isolated units; they exchange energy and materials with surrounding ecosystems.
- Similarly, urban ecosystems exchange energy and materials with surrounding ecosystems. Cities are not self-supporting.
- The effect of cities is greatest on their immediate surroundings, but, due to modern transportation networks, cities may draw upon resources from virtually anywhere on the globe.
- However, by concentrating human populations in properly planned cities, the effects on the local and regional landscape can be reduced and land that would be occupied by urban sprawl can be kept open as green space or for other uses.

Historically, what factors have determined the location of cities?

- Two factors usually determine city locations: site and situation.
- **Site** is the combination of environmental features of a location. Site is usually determined by geologic and soil conditions, water supply and hydrology, agricultural potential, the availability and accessibility of resources, and climatic factors.
- **Situation** is the location of a city in relation to other cities or areas. Situation is primarily dependent upon accessibility to transportation networks (waterways, roadways, aviation, and rail lines). Historically, cities were sited in areas that could be suitably defended against aggressors, but this aspect of their situation is less important today.
- The rise of the modern industrial metropolis was made possible only after two basic technological advancements: improved transportation and improved sanitation.
- Most of the world's great cities are located where there is a combination of good site and good situation. Typically, these cities are located in coastal areas with good harbors or adjacent to navigable rivers.
- San Francisco, Los Angeles, New York City, Chicago, Seattle, Istanbul, Athens, Bombay, Singapore, Hong Kong, and Buenos Aires are all port cities with good harbors.
- St. Louis, London, Cairo, Paris, Moscow, Beijing, and New Delhi are located on major rivers.
- In the eastern and southeastern United States many cities and towns are located along the "fall line" where the piedmont plateau adjoins the coastal plain (Figure 27.5); these locations provided access to rivers for transportation and hydropower, as well as good agricultural lands.
- In the absence of good site conditions, site modifications were made if the city situation was justified. New Orleans has poor site quality (low-lying and flood-prone) but its situation at the mouth of the Mississippi justified the construction of massive flood control systems. Portions of cities such as Washington, San Francisco,

Oakland, and New York were located in low-lying tidal areas that were filled as the cities expanded.

- If the adverse environmental effects of cities are too great and site quality declines, situation may decline as well.
- Many cities have outgrown their support systems. A common example is water supply. As much as one hundred years ago, New York and San Francisco had to develop water supplies in far removed watersheds and pipe these waters over great distances. Decades ago, Los Angeles, Las Vegas, and Phoenix outgrew local water supplies and became reliant on distant sources. Many rapidly growing urban areas in the southeast, such as Norfolk-Virginia Beach, are experiencing critical water supply problems. In an ecological sense, water should be a limiting factor of growth; but local decision-makers rarely consider restricting growth and water supplies are developed to the detriment of the environment.
- Today, many cities grow in response to employment opportunities. The microelectronics industry of California's "Silicon Valley" and the high technology and biotechnology industries of the Research Triangle Park in North Carolina are the reasons cities in these areas have experienced dramatic increases in population.
- As the "baby boom generation" enters retirement age, retirement communities in moderate climates are expected to grow.

Historically, what has been the role of city planning in the urban environment?

- Cities that have developed with little or inappropriate planning quickly lose any natural attributes that may have existed, may become impoverished, congested, and polluted. Environmental quality and quality of life will decline. Several cities in developing nations exhibit these conditions, such as Mexico City, Calcutta, Jakarta, and Sao Paulo. As cities decline, the surrounding landscape becomes degraded as well.
- Well-planned cities usually remain economically viable with higher standards of living and higher quality of life. Well-planned cites are characterized by:
  - accessibility to necessary resources
  - well functioning transportation systems (good roads and thoroughfares, mass transit systems, relatively easy access to areas within and beyond the city),
  - pollution control and waste disposal systems that are adequate for the volumes that must be treated,
  - parks, natural areas, and open areas that provide refuge and buffers from the artificial landscape.
- Some city planners recognize the long-term value of greenbelts and other more environmentally sound aspects of development. Unfortunately, many cities opt for short-term economic gain at the expense of environmental sustainability. The prevalence of strip malls in almost every urban area of the United States is evidence of this.

What are some of the environmental risks urban dwellers encounter?

- Compared to people living outside of cities, people who live in cities are usually exposed to higher concentrations of pollutants and for longer durations.

- Increased exposure to toxic chemicals, air pollutants, heat and noise contribute to increased stress and health risks in urban areas of the United States.

What are some of the effects of cities on local climate, hydrology, and soils?

- Cities modify local climate (discussed in Chapter 22).
  - Compared to the countryside, cities are hotter (heat islands develop) (Figure 27.11), drier air, higher precipitation, and have less wind to disperse air pollutants.
- Cities concentrate air pollutants (discussed in Chapter 23).
  - Stationary and mobile sources of air pollutants are more numerous in cities.
- Cities greatly affect local and regional hydrology.
  - Rainfall may be increased because urban dust provides condensation nuclei for water droplet formation, and cloud and fog formation.
  - Impervious surfaces (paved roads and parking lots, and rooftops) do not allow precipitation to infiltrate into the soil; the resulting increased runoff increases the chances and intensity of flooding downstream.
  - Levees are built along riverbanks to protect cities from flooding funnel floodwaters downstream where flooding in rural and agricultural areas or neighboring towns becomes more severe. Natural floodplains are decoupled from rivers and sediments are transported downstream.
  - In many cities, sewage systems are not capable of the increased volume of water received during floods; consequently raw sewage may be discharged into rivers or bays.
  - Most cities do not provide basic wastewater treatment for stormwater runoff. Stormwater, which contains litter (e.g., plastic bottles and bags and paper), oils, gasoline, grease, and solvent from automobiles, animal feces, and an array of other contaminants, is discharged untreated directly to receiving waters.
- Cities affect soils and soil erosion.
  - Natural soils are dynamic systems where geologic, hydrologic, and biologic interactions take place. Paving, covering, compacting, or building on soils causes these interactions to cease.
  - Soils from construction sites are easily eroded and sediments may enter streams and riverbeds.

How can the urban environment be designed to increase environmental quality and quality of life?

- The cities that are often considered the most aesthetically pleasing cities usually have a large number of parks, greenways, natural areas, attractive waterfronts, and historical sites.
  - The history of urban parks and urban environmental planning dates back to the Mediterranean cultures of the first millennium A.D.
  - The more modern era of park design was typified by the work of landscape architects such as Frederick Law Olmstead and Calvert Vaux, the designers of Central Park in New York. Olmstead also designed a watercourse to control flooding and treat wastewater in Boston.

- Boston has the Minuteman Trail (a greenway/bikeway), and numerous old cemeteries, such as Mount Auburn Cemetery, which is as interesting horticulturally as it is historically.
- Charleston, South Carolina has numerous privately owned gardens and small parks.
- A focal point of Washington is the Potomac River, which until recently was severely polluted but now is an important recreational boating resource for kayakers.
- Mt. Rainier, which dominates the Seattle skyline, is a constant reminder of the natural beauty of the Pacific Northwest.
- Many of the U.S. cities that experienced urban decay in the 1960s and 1970s successfully focused revitalization efforts on parks and waterfronts (e.g., Baltimore and Detroit).

- Trees have been used to enhance the beauty and aesthetic qualities of cities, like Paris, for centuries. Other benefits of urban forestry and trees include modifying microclimate, absorbing air pollutants, providing noise buffers, and providing habitat for urban wildlife.
- Planting deciduous trees on the south side of houses provides summer shade and winter sunlight. Conifers planted on the north side of houses provide some insulation from cold winter winds.
- Cities are stressful environments for trees. Air pollution, soil compaction, sidewalks, and physical damage tend to increase susceptibility of trees to disease and insect attack, and stunt growth and shorten the life span of trees. Trees species vary in their suitability to urban environments.
- Suitable habitats for some types of wildlife can be provided by cities. Wildlife species can either not tolerate urban environments and disappear, marginally tolerate urban environments, or thrive in urban environments where their relationship with humans is either neutral, beneficial, or as pests.
- Urban food chains based on plants and insects or on road kills can develop. Endangered peregrine falcons have found a niche in New York City. These birds of prey have been observed foraging on urban pigeons and nesting on the ledges of high rise buildings.
- Pest infestations in cities can be significant health risks. Under the right conditions, pests such as cockroaches, fleas, rats, and pigeons can be disease vectors.

Ecology In Your Backyard

- Explore the greenbelts and natural areas of your community or campus. What wildlife do you see, what plant species are present? Urban areas contain many non-native species of plants, birds, and other animals. Do you know if the species you observe are native or non-native species?

- As you explore your urban environment from an ecological perspective, what environmental problems are evident? Are there soil erosion problems or trees in poor health? Take a trash bag along, any litter that you remove from streets and stormwater grates is litter that is not flushed downstream.
- Many songbird species thrive in parks, lawns, and residential areas of the urban environment. However, the urban environment may also be quite hazardous. Domestic animals (cats and dogs) can be effective predators. Buildings and roads are also great hazards to birds and other urban wildlife. Glass windows, bright lights, and turbulent winds of building kill thousands of birds each year. Make a monthly check around the perimeter of a high-rise building in your community or campus and record the mortality rate.

  Highways and roads inflict a high mortality on species wildlife as well as domestic animals. If safety allows, keep a record of road kills (by taxonomic group) that you spot while traveling a regularly traveled route. You might be surprised at the numbers.

Links In The Library

- Aberly, D. ed. 1994. *Futures By Design: The Practice of Ecological Planning.* New Society Publishers, Philadelphia.
- *Environmental and Urban Issues.* A journal published by Florida Atlantic University/Florida International University - Joint Center for Environmental and Urban Problems.
- Lowe, M. 1992. City limits. *World Watch* 5 (1): 18-25.
- Odum, H.T. 1971. *Environment, Power, and Society.* John Wiley and Sons, New York. 331 pp.
- Population Action International. Cities: Life in the World's 100 Largest Metropolitan Areas.

Ecotest

1. In the U.S., _____ % of the population lives in urban areas.

a. 25
b. 55
c. 65
d. 75

2. The location of cities has traditionally been determined by site and _________________.

a. demographics
b. socioeconomics
c. situation
d. resources

3. The principal designer of Central Park in New York City was _________________.
a. Fuller
b. Olmstead
c. Wright
d. Tweed

4. Cities are sometimes referred to as ____ because they tend to trap and retain heat.
a. heat pyramids
b. heat exchangers
c. heat shields
d. heat islands

5. Cities have all of the following effects on local hydrology except:
a. decreased rainfall.
b. increased runoff.
c. decreased infiltration.
d. increase the severity and frequency of downstream flooding.

6. Many cities in the eastern and southeastern U.S. are located along the _________ where rivers flow out of the piedmont and into the coastal plain.
a. seaboard
b. plateau
c. fall line
d. estuaries

7. According to your text, two technical advances made possible the rise of the modern industrial metropolis, improved transportation methods and _______________.
a. improved sanitation methods
b. improved food delivery
c. improved energy availability
d. improved water supplies

8. Which endangered bird of prey has been observed nesting on ledges and foraging on pigeons in New York City?
a. golden eagles
b. caracaras
c. peregrine falcons
d. king vultures

9. In the U.S., summer shade and winter sunlight for houses can be provided by trees if :
a. deciduous trees are planted on the south side of the house.
b. deciduous trees are planted on the north side of the house.
c. coniferous trees are planted on the south side of the house.
d. coniferous trees are planted on the north side of the house.

10. Today, an estimated ____% of the global population will be living in urban areas.
a. 25
b. 45
c. 75
d. 85

11. How many cities are expected to have populations over 8 million by the year 2015?
a. 6
b. 12
c. 36
d. 48

12. Cities are often located along the "fall line" because they benefit from:
a. good agricultural land nearby.
b. easy access to navigable portions of rivers.
c. readily available source of hydropower.
d. all of these are possible reasons.

13. Well-planned cities are characterized by all of the following except:
a. accessibility to natural resources.
b. well functioning transportation systems.
c. pollution control mechanisms.
d. limited access to areas within the central city.

14. Urban environments act as an ecological system in all of the following ways except:
a. they both exchange energy and materials with other systems.
b. they both are biotic communities.
c. they both undergo successional changes with time.
d. they both are entirely self-supporting.

15. All of the following are site factors that determine a city's location except:
a. climatic factors.
b. water supply and hydrology.
c. the proximity of the nearest airport.
d. suitability of the soil and underlying geology.

Chapter 28 - WASTE MANAGEMENT

The Big Picture

Two factors are fundamental to understanding the problem of waste disposal. First, the burgeoning human population consumes an increasingly large portion of our natural resources and consequently generates more waste materials. Second, the waste materials generated are often not easily biodegraded in the environment. Traditional methods of waste disposal are no longer adequate and alternatives must be pursued. Land-filling is the most commonly used means of solid waste disposal in the United States, but landfills are problematic. Landfills require large areas with the appropriate site conditions. Such areas are increasingly difficult to find. Even the best designed landfills may ultimately develop leaks and leachate could contaminate groundwater or surface water. Hazardous waste disposal presents additional problems. Hazardous wastes constitute less volume than solid wastes, but pose a greater risk to the environment and human health. Disposal sites for hazardous wastes must be designed to prevent the movement of contaminants off-site and rigorously monitored to ensure groundwater, surface water, and air are adequately protected. The implementation of Integrated Waste Management (IWM) strategies may be a partial solution to solid and hazardous waste disposal problems. Basic to IWM is the reduction of the amount of waste generated and reusing or recycling waste materials.

Frequently Asked Questions

What are solid wastes and hazardous wastes?

- Solid wastes consist of nontoxic materials discarded from households, industry, and agricultural sectors.
- Waste materials vary dependent upon local land use, economic base, industrial activity, climate, and season.
- In the United States, paper products constitute the bulk of the solid waste that is deposited in landfills (Figure 28.2).
- In the United States, newspaper constitutes as much as 18% of the total waste volume disposed of in urban landfills.
- Hazardous wastes are materials that present an acute or chronic risk to human health. This includes toxic (e.g., pesticides, heavy metals, solvents) and flammable liquids (e.g., petroleum products, paints) (Table 28.1).

How have attitudes toward wastes changed?

- In the past, as long as human populations remained small and the waste materials they generated were relatively easy to decompose in the environment, uncontrolled dumping of waste into rivers, bays, or out-of-the-way places in the landscape did not cause widespread problems. This exemplifies the concept of "dilution and dispersion."

- Today, human populations are large and mostly concentrated in urban areas. The waste materials generated are of much greater volume and often contain high concentrations of biopersistent hazardous materials. Waste material cannot be cast, untreated, into the landscape without serious environmental and human health degradation. Most wastes are thus disposed of by "concentration and containment" methodologies, especially landfilling.
- The environmental catastrophe at Love Canal, New York in 1976 initiated nationwide concern about the risks that hazardous and toxic chemicals present to human health and to the environment.
- A recent trend in waste management is the utilization of a variety of options known collectively as "integrated waste management" (IWM).

What is integrated waste management (IWM)?

- Instead of relying primarily on concentration and containment technologies of waste disposal, IWM offers a variety of waste management alternatives.
- IWM alternatives include: reducing the amount of waste generated, reusing "waste" materials when possible, recycling "waste" materials, composting organics materials, as well as the more traditional waste disposal technologies of landfilling and incineration.
- If implemented, IWM could reduce the amount of waste disposed of in landfill and incinerators by 50% to 70%.

Does recycling really work?

- In the United States, a 30% reduction of urban waste materials through recycling programs is a reasonable goal. Although, the potential for waste reduction is much higher.
- The U.S. recycling rate was 27% in 1996.
- Generally, there is strong public support for recycling programs.
- A major limitation for successful recycling programs is the lack of markets for recycled materials.
- Recycling should be considered a third option for disposing of waste materials; reduction and reuse options should be considered first.
- Materials management is a new approach to waste reduction that would eliminate the harvest of virgin natural resources, establish incentives to use recycled materials in the construction of new buildings, establish penalties for negative materials management practices, provide financial incentives for environmentally sustainable industrial practices, and increase the number of new jobs associated with materials management and recycling.

Can human waste be recycled?

- Historically, human waste has been used as fertilizer in many cultures for centuries.
- Human waste is still a potential fertilizer source, even in highly urbanized cultures, given the proper conditions.

- The waste must be free of or at least contain low concentrations on pathogens or hazardous chemicals.
- Waste from municipal wastewater treatment plants is generally the least suitable but waste from small-scale and localized treatment systems has greater promise. Of course, one of the biggest drawback is one of public perception.

What is materials management?

- Materials management is a general term used to describe more efficient use of materials in an effort to reduce the amount of material generated that could ultimately become waste material.
- Materials management entails eliminating subsidies for raw materials, encouraging the use of recycled materials, assessing financial penalties for improper material management practices, providing financial incentives for environmentally sound materials management practices and encouraging new technologies that generate new jobs in the reuse and recycling of materials.

What is a sanitary landfill?

- Sanitary landfills are waste disposal sites that utilize compaction and containment technologies.
- Sanitary landfills are engineered to minimize the movement of leachate (contaminated water) off-site via groundwater or surface water.
- The best designed landfills are situated in areas with a clay-dominated subsoil. The rate of leachate movement through clay soils is very slow.
- Plastic or other impermeable liners may be used to further impede the infiltration of leachate into groundwater.
- To reduce rainwater infiltration from above, a compacted clay cap is placed on top of newly deposited wastes at daily intervals.
- Monitoring wells are established in key locations around the landfill and groundwater quality is recorded.
- Social concerns over landfill site selection have to be addressed as well as topography, geology, hydrology, and climate.

What environmental and human health risks are associated with landfills?

- Access to landfilled wastes by animals, despite daily capping and prevention measures.
- Atmospheric inputs of gases produced by decomposing wastes.
- Heavy metal contamination of the soil.
- Groundwater contamination with soluble materials, thus the potential contamination of drinking water supplies.
- Contamination of overland runoff which may enter streams and rivers.
- Uptake of contaminants by plants growing in waste disposal sites and subsequent transfer through food webs.

How are landfills regulated?

- Landfills must be in compliance with federal, state, and local regulations during the site selection, design, operational, and monitoring phases.

- State and local regulations will obviously vary, however, federal regulations, promulgated by the Resource Conservation and Recovery Act (RCRA), help to ensure that nationwide standards are maintained.
- RCRA is administered by the U.S. Environmental Protection Agency (EPA). RCRA offers states an option of rigid compliance of federal regulations or more flexible landfill plans if approved by EPA and meeting minimum standards.
- Nationwide standards include:
    - avoidance of floodplains, wetlands, earthquake faults, unstable land, and sites near airports
    - landfill liners
    - leachate collection systems
    - groundwater monitoring
    - financial assurance that funds are available for continued landfills monitoring

How extensive is the problem of uncontrolled hazardous waste disposal?

- The United States generates over 150 metric tons of hazardous chemical wastes, annually.
- Past and present uncontrolled dumping has resulted in an estimated 32,000 to 50,000 dump sites in the United States. Of these, 1,200 to 2,000 pose a serious threat to public health or environmental quality.
- Improperly stored barrels of hazardous wastes may develop leaks and contaminate soil, groundwater, and surface water.
- Unlined lagoons and improperly constructed landfills may allow hazardous liquids to infiltrate into groundwater.
- Illegal dumping in deserted areas, along roadways, and in landfills that are not designed to accept hazardous wastes may result in soil and water contamination.

What roles do RCRA and CERCLA have in regards to hazardous waste disposal?

- The Resource Conservation and Recovery Act (RCRA) is the "cradle-to-grave" legislation designed to track hazardous wastes from their generation to their ultimate disposal. The goal is to ensure that the hazardous wastes are properly stored and disposed of.
- The Comprehensive Environmental Response Compensation and Liability Act (CERCLA) established policy and procedures for hazardous waste disposal. CERCLA also required a nationwide listing of hazardous waste sites with the most serious contaminant problems, the National Priorities List (NPL).
- NPL sites have been targeted for site remediation or clean-up; this may be very costly. Because many NPL waste sites are abandoned or the site owners are unable to pay, CERCLA established a special fund, Superfund, to cover clean-up costs.
- To date, relatively few sites have been completely remediated through the Superfund program.

What are the conventional methods of disposal for hazardous wastes?

- Secure landfills are similar to sanitary landfills with the exception that additional precautions are taken to prevent the movement of leachate into groundwater (Figure

28.10). Siting considerations are important, however, almost all landfills are subject to leakage. At secure landfills, leachate is collected and treated.

- Surface impoundments are designed to contain liquid hazardous wastes. These impoundments are especially vulnerable to leakage. Aerosol pollutants may also emanate from impoundments.
- Deep-well disposal involves injecting hazardous wastes into subterranean permeable rock layers or abandoned mine sites. This method assumes that these geologic formations are stable and do not come in contact with aquifers. Hazardous wastes disposed of in this manner cannot be retrieved for additional treatment.
- Land application is the application of wastes onto land surfaces thus promoting decomposition by aerobic microbes. Land application is effective for biodegradable, organic wastes only. If not sited properly, surface runoff may become a problem.

Can the concepts of IWM be adapted to hazardous waste disposal?

- As with solid wastes, IWM concepts could be implemented.
- Source reduction can reduce the amount of hazardous waste generated.
- Recycling and recovery can be used to generate new products from hazardous waste.
- Hazardous wastes can be treated to reduce their toxicity or hazardous characteristics.
- Hazardous wastes can be incinerated and the less toxic or hazardous by-products of incineration can be treated.

What problems are associated with ocean dumping of solid and hazardous waste?

- Ocean dumping harks back to the days of waste disposal by "dilution and dispersion." Certainly, the huge volume of the world's oceans is capable of dilution and dispersion, but pollutants are not evenly dispersed throughout the entire volume.
- Near-shore areas and localized sites in the vicinity of waste discharge outfalls or dump sites have become seriously degraded.
- The United States has reduced but not eliminated the amount of waste discharged into the oceans. Internationally, ocean dumping is ongoing and consensus among nations that participate in ocean dumping is problematic.
- Dredge spoils (sediments excavated from bay, channels, and canals to improve navigation) constitute the largest volume of wastes disposed of by ocean dumping. Potentially more hazardous and toxic industrial waste, sewage sludge, construction and demolition debris, and solid waste are also disposed of in the oceans.
- The pollution of the oceans as a result of ocean dumping results in the exposure of marine organisms to toxic materials which may result in retarded growth, vitality, reproductive ability, or death. Water quality is degraded by a lowering of dissolved oxygen concentrations and increased eutrophication. The consequences are a loss of the more sensitive species in the marine ecosystem. The reduction in abundance or complete loss of any species in an ecosystem may result in significant disruptions to the entire system.

How have waste management approaches changed from focusing on treatment of waste once generated to a focusing on pollution prevention?

- By the 1990's there was a shift from a focus on treatment of waste already generated toward a focus on reducing the amount generated in the first place as well as reducing the degree of hazard of that waste. Thus waste treatment and pollution prevention has becoming more proactive than reactive.
- Pollution prevention approaches include reducing excessive use of raw materials, more efficient use in manufacturing products, using nontoxic materials instead of toxic materials.

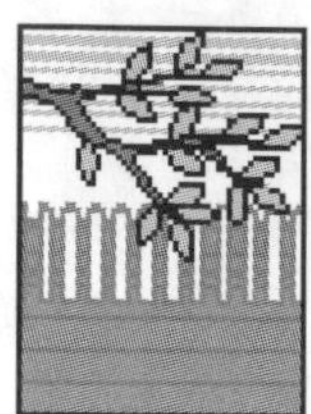

Ecology In Your Backyard

- Does your college, university, or community have a recycling program?
- If so, is it effective?
- Are collection sites readily accessible and well used?
- If the program is effective, you may consider showing your appreciation to the administration for a job well done.
- If not, query the administration and suggest improvements.
- Composting may not be possible on an individual basis in some urban areas. However, if you are in a situation in which composting is an option and the composted material can be used, find out how to safely compost organic wastes.
- Consider the products that you buy at the grocery store or department store; this is where you can play a direct and immediate role in waste reduction. Does the amount and type of packaging affect your choice of a product? Does buying in volume help to eliminate some packaging wastes?

Links In The Library

- Bailey, J. 1995. The Recycling Myth. *The Wall Street Journal*. January 19, 1995. This article calls into question the environmental benefits of recycling. This author suggests that landfill space is plentiful, recycling causes more pollution than it prevents, and recycling costs cities more money than it saves them. But are all the costs and savings included?
- Evans, D. G. 1994. A rationale for recycling. *Environmental Management*. Vol. 18 No. 3 - This author argues that even though market forces may not make recycling profitable, we should expand recycling efforts with government subsidies, because it would achieve a desirable goal of our society.
- Young, J. E. 1995. The sudden new strength of recycling. *World Watch*, July/August 1995. - This article details the cost savings that certain cities gained when they instituted recycling programs.

- Hendrickson, C., L. Lave, and F. McMichael . 1995. Time to dump recycling? Issues in *Science and Technology*. Spring 1995.

Ecotest

1. Which of the following waste categories comprises the largest percentage of solid waste deposited in landfills?
a. paper
b. glass
c. plastics
d. old appliances

2. The three "R's" of a well-designed waste management plan stand for:
a. refine, remediate, recycle.
b. respond, recover, restore.
c. recover, recycle, remediate.
d. reduce, reuse, recycle.

3. Leachate from landfills migrates slowest through _____, so landfills should have a liner of this soil type.
a. sand
b. silt
c. clay
d. gravel

4. Nationwide standards for landfills were established by which federal legislation?
a. CERCLA (Comprehensive Environmental Response, Compensation, and Liability Act)
b. RCRA (Resource Conservation and Recovery Act)
c. NEPA (National Environmental Policy Act)
d. GWPA (Groundwater Protection Act)

5. Which federal legislation provides financial and technical assistance to offset costs associated with the clean-up of sites that have been severely contaminated by hazardous and toxic materials.
a. CERCLA (Comprehensive Environmental Response, Compensation, and Liability Act)
b. RCRA (Resource Conservation and Recovery Act)
c. NEPA (National Environmental Policy Act)
d. GWPA (Groundwater Protection Act)

6. The United States generates approximately ____ metric tons of hazardous chemical waste annually.
a. 150

b. 300
c. 375
d. 425

7. How do secure landfills differ from conventional landfills?
a. Secure landfills are encased in asphalt or concrete to eliminate the possibility of leachate migration.
b. Secure landfills may be sited only on federal lands in remote, arid regions of the United States.
c. Secure landfills collect and treat leachate.
d. Secure landfills use high security fencing and motion detection devices to inhibit site access by humans and animals.

8. By volume, which of the following waste categories comprises the bulk of waste dumped by the United States in oceans?
a. dredge spoils
b. municipal garbage
c. abandoned or dismantled transportation infrastructure (e.g., dismantled bridges, roads, railroad cars)
d. military equipment (e.g., scuttled ships, discontinued armament and transport vehicles)

9. Which of the following is not a viable option for hazardous waste disposal?
a. implementing the concepts of Integrated Waste Management (IWM)
b. incineration
c. deep-well injection
d. ocean dumping

10. In the United States, the environmental and human health problems associated with hazardous waste disposal first gained national attention with the incidents at _____________.
a. St. Lucie River, Florida
b. Chemical Alley, Louisiana
c. San Francisco Bay, California
d. Love Canal, Niagara Falls, New York

11. In the United States, newspaper constitutes as much as _______________ of the total waste volume disposed of in urban landfills.
a. 10 %
b. 18 %
c. 25 %
d. 48 %

12. Sediments excavated from bays, channels and canals to improve navigation are called:
a. epibenthic fauna.
b. dredged spoil.
c. sludge.

d. overburden.

13. Which method of waste disposal assumes that subterranean geologic formations are stable and therefore acceptable sites for waste disposal?
a. sanitary landfills
b. surface impoundments
c. land application
d. deep-well disposal

14. In the past, as long as human populations remained small and the waste materials they generated were relatively easy to decompose in the environment, uncontrolled dumping of waste into rivers, bays, or out-of-the-way places in the landscape did not cause widespread problems. This exemplifies the concept of:
a. "dilution and destruction."
b. "dilution and dispersion."
c. "destruction and mayhem."
d. "the solution to disposal."

15. Which of the following is not a hazardous waste?
a. animal manure
b. hypodermic syringes
c. lead-based paint
d. fuel-grade uranium from a nuclear reactor

Chapter 29 - MINERALS AND THE ENVIRONMENT

The Big Picture

When you stop to consider all of the ways we use minerals, you begin to realize the importance of minerals and mineral usage in our technological society. A typical classroom building on a university campus is constructed almost entirely of minerals. The bricks, blocks, mortar, glass, steel beams, copper wiring, sheet rock, paints, and insulation are all made from minerals that have been mined, processed, and manufactured into their present forms. Roads, bridges, buildings, trains, aircraft, and other infrastructure are largely made of minerals. Your automobile, furniture, appliances, and computer contain components made from minerals. The plants that provide you with food and clothing (natural fibers), probably require mineral-based fertilizers. Geologists recognize that mineral deposits are very unevenly distributed on and within the Earth's crust and oceans. In order for minerals to be mined economically, deposits must be of sufficient concentration to allow extraction with current technologies. Mineral exploration, mining, and processing may have adverse environmental effects, as well as socioeconomic effects on mining communities. By improving exploration, mining, and processing technologies, requiring mine site reclamation, and adherence to international, national, state, and local laws, environmental degradation can be reduced. Mineral reserves and resources are finite. If the production of mineral-based goods is to continue, efforts must be made to treat minerals as a finite resource and apply the concepts of reduction, reuse, and recycling to minerals.

Frequently Asked Questions

What are minerals?

- Minerals are inorganic materials that have a definite internal structure and that have specific physical and chemical properties.
- Minerals include non-metals (e.g., sulfur, phosphate, silica) and metals (e.g., iron, lead, titanium).

What are minerals used for?

- Certain minerals are essential for life. For example, green plants uptake phosphate (a structural component of DNA and ATP) from the soil. Phosphate may be transferred to consumers via food webs.
- Minerals are also used in the manufacture of almost all infrastructure and commodities used in modern technological cultures (Table 29.1).

How are mineral deposits formed?

- The elements, which constitute minerals, are unevenly distributed in the Earth's crust and oceans. Theoretically, as the proto-Earth cooled, lighter weight minerals formed the crust while heavier metals (principally iron) formed the earth's core. Tectonic activity of the Earth's crust and mantle subsequently redistributed minerals (Figure 4.3).
- Of the 112 known elements (Figure 4.5), only eight comprise 99% of the Earth's crust. In decreasing order of abundance, they are: oxygen (O), silicon (Si), aluminum (Al), iron (Fe), calcium (Ca), sodium (Na), potassium (K), and titanium (Ti).
- Dissolved solids (minerals) constitute about 3.5% of the weight of seawater. Most of the dissolved mineral is chlorine, but smaller concentrations of other non-metals and metals may be found. The total salts dissolved in seawater average 35 parts per thousand (ppt).
- Metals and minerals are concentrated by igneous and sedimentary processes.
- Gold ore is typically found in association with the igneous rock, quartz.
- Sedimentary processes, such as the weathering and erosion of gold ores, may result in the deposition of gold in placer deposits. Placer deposits in California and Alaska were extensively mined in the 19th century and early 20th century.
- Mineral deposits can be formed through evaporative processes. As shallow seas evaporate, marine evaporites such as potassium and sodium salts, gypsum, and anhydrite accumulate on former sea beds.
- Similarly, on inland salt lakes, near thermal springs, or other brine waters, evaporites such as borate, iodine, calcium chloride, and magnesium accumulate.
- Mineral deposits can form as a result of biological processes. Ancient organisms may have indirectly influenced the deposition of iron ore by altering the concentration of oxygen in the atmosphere. Phosphate and calcium deposits formed in shallow seas where the bones and shells of marine organisms accumulated.
- Secondary enrichment occurs when acid groundwater leaches minerals from primary ores and redeposits them at or below the water table. Minerals in the enriched zone may be as much as ten times more concentrated thus more economical to extract.
- The mineral resources of the oceans are vast. Presently, most of these resources (with the exception of magnesium) are not economically feasible or technically possible to exploit. Minerals associated with sulfide deposits at thermal vents (black smokers; Figure 29.3) and manganese oxide nodules on the sea floor may become economically and technologically feasible to mine in the future when extraction and metallurgical processes improve.

What are mineral resources and mineral reserves?

- Mineral resources are mineral deposits that are presumed to exist in sufficient concentrations to be economically exploited (Figure 29.4).
- Mineral reserves are a sub-category of mineral resources that have been surveyed and determined to be economically feasible to exploit.

- As new deposits are discovered, extraction technology improves, or economic changes increase the value of deposits, mineral resources may be converted to mineral reserves.
- Mineral resources are finite. Before mineral resources are exhausted completely, however, the cost of exploitation will exceed the value of the mineral. When this situation occurs, the options are:
  1) find more resources;
  2) recycle what has already been obtained;
  3) find a substitute; or
  4) do without.
- Historically, the patterns of resource consumption that have been observed are: 1) rapid consumption, which leads to resource depletion; or 2) conservation, and 3) a combination of conservation and recycling which tend to maintain the resource at a more sustainable level (Figure 29.5).

What is the status of the mineral resources for the United States?

- The United States along with the other developed nations account for only 16% of the global human population but consume a disproportionate share of the mineral reserves. For example, the U.S., western Europe, and Japan consume 70% of the total aluminum, copper, and nickel that is mined.
- Domestic supplies of some minerals in the United States are not sufficient. Even though substantial resources may exist within the jurisdiction of the United States, economic, political, or environmental reasons may justify reliance on foreign suppliers.
- Federal lands, especially in the western United States, contain extensive mineral resources that may be exploited if future demand warrants.
- Relying on imported minerals is less secure than relying on domestic reserves and resources. Political, economic, or military instability in foreign supplier nations may jeopardize United States imports.
- Strategic minerals are those that are essential for military or industrial application. The United States maintains stockpiles of many strategic minerals for reasons of national security.
- As with most environmental problems, the demand for and depletion of mineral reserves and resources is exacerbated by increasing human population.

What are some of the environmental effects of mineral exploitation?

- Mineral exploitation has three phases: exploration, mining, and processing. The environmental impacts of exploitation vary depending upon the phase and conditions at the mining site.
- Mining and processing usually have more adverse environmental impacts than exploration.
- Site conditions, such as ore quality, local hydrologic conditions, climate, rock type, size of operation, topography, and other factors, result in varying environmental impacts among mining sites.

- Exploration activities, such as road building and vehicular traffic, may have adverse environmental impacts in especially sensitive areas such as tundra, desert, and wetlands.
- Mining activities are especially damaging if low-grade ores are being extracted, thus necessitating the removal of large volumes of ore.
- Surface mining (open-pit mining) is cheaper but usually more damaging than subsurface mining. Surface mining involves the obliteration of the ecosystems that lie atop the ore body. Local and regional aquifers may be drastically dewatered by surface mining.
- The processing of mineral ores generates mine tailings (wastes) that may contain hazardous, toxic, or acidic materials. Trace elements of toxic heavy metals (cadmium, cobalt, copper, lead) may leach into drinking water supplies or drain into streams.
- For some minerals, sulfur compounds are used to separate the mineral from its ore. The resulting acid mine drainage from this process can result in stream and soil acidification.
- Sulfur emissions can contribute to air pollution in the vicinity of mines, thus damaging plants, animals, and affecting human health.
- The federal Surface Mine Reclamation Act of 1977 requires the reclamation of coal mine sites. Mine sites from which minerals are extracted are not required to reclaim the site under federal authority, however, many states require reclamation and monitoring.
- Rarely, if ever, do reclamation projects attempt to restore sites to pre-mining conditions. Instead, the sites are backfilled and planted in a cover crop or flooded.

What are some of the social impacts of mineral exploitation?

- When mineral exploitation is initiated, local social structure may be affected. New mining operations result in an influx of workers, which may stress local infrastructure. Land use patterns change as rural landscapes are converted to industrial and urban landscapes. Water quality and air quality are adversely affected by direct and indirect consequences of mining.
- The economic base of the local community may shift toward greater reliance on the mining operation. When the mine is no longer profitable to operate, the mine closes. Unless economic alternatives are found, communities are faced with the loss of a major contributor to the local tax base and extensive unemployment. The 19th century boom town to ghost town phenomenon in parts of the western United States was a testament to the impact of mining on the socioeconomic structure of communities.

Mineral exploitation is a necessity. Mineral exploration, mining, and processing will continue to alter landscapes and generate pollutants. What can be done to minimize the adverse environmental impacts of mineral exploitation?

- If the adverse environmental impacts of mineral exploitation are to be minimized, action must be taken at each step of the resource cycle (Figure 29.8)

- Legislators and regulators at the federal, state, and local levels must continue to enact, support, and enforce regulations that address the pollution problems arising from mining activities.
- Mining wastes must be treated both on-site and off-site to prevent sedimentation problems in streams, water pollution and air pollution. Innovative engineering and biotechnological solutions should be explored.
- Practice the three R's of waste management. Reduce the production of mining wastes, reuse and recycle minerals and their wastes.

Ecology In Your Backyard

- What minerals are mined in your area?
- Where are the mines located and what (if any) environmental problems have resulted from the mining operations?
- Is the mining company currently reclaiming any mined sites?
- How do you think the community perceives the mining operation?

Links In The Library

- Broadus, J.M. 1981. Seabed materials. *Science* 235(4791): 853-860.
- Brown, L. et. al. Mining the Earth. In *State of the World, 1992*. W.W. Norton and Company, New York.
- Cameron, E.N. 1986. A*t the Crossroads: The Mineral Problems of the United States*. John Wiley and Sons, Inc., New York.
- Dubs, M. 1986. Minerals of the Deep Sea: Politics and Economics in Conflict Pages 55-83 IN *Ocean Yearbook 6* Borgese, E.M. and N. Ginsburg, eds. University of Chicago.
- Frosch, Robert A. and Nicholas E. Gallopoulous. 1990. Strategies for manufacturing. Chapter 9 in Managing Planet Earth: readings from *Scientific American Magazine*. W. H. Freeman and Company, New York. This article discusses the use of various minerals in the manufacturing process, estimated lifetimes of various minerals given current rates of use, and given predicted rates of use by the year 2030. Some minerals will become very scarce if these predictions are true.

Ecotest

1. Eight elements constitute 99% of the earth's crust. Which of the following is not one of those elements?
a. silicon
b. aluminum
c. oxygen
d. nitrogen

2. The average salinity of the oceans is?
a. 35 %
b. 35 ppt
c. 35 g/ml
d. 35 g/cm^3

3. Which of the following statements about mineral reserves and resources is false?
a. Mineral resources are a sub-category of mineral reserves.
b. Mineral reserves and resources are finite.
c. Mineral reserves and resources are unevenly distributed, globally.
d. Mineral resources may become mineral reserves depending upon the cost of extraction.

4. The Surface Reclamation Act of 1977 requires site restoration upon the completion of mining for ________.
a. all minerals
b. non-metals
c. radioactive minerals
d. coal

5. The depositon of ____________ is a consequence of biological processes, as well as geologic processes.
a. phosphate
b. silicon
c. titanium
d. aluminum

6. The United States, western Europe, and Japan collectively consume ___ percent of the aluminum, copper, and nickel that is extracted worldwide.
a. 24
b. 30
c. 55
d. 70

7. Minerals may be deposited near oceanic vents that are ejecting hot, sulfide-rich water called ______.
a. fumerols
b. lava domes
c. black smokers
d. enrichment zones

8. Many streams in the coal mining regions of the Appalachian Mountains and the Allegheny Plateau have been contaminated by ______ which develops as rainwater runoff flows through mining sites and mining wastes.
a. silicon dioxide
b. acid mine drainage
c. acid rock
d. acid precipitation

9. Certain minerals that are stockpiled by nations for reasons of national security (military and industrial) are called _________.
a. strategic minerals
b. essential minerals
c. imperative minerals
d. fundamental minerals

10. All of the socioeconomic effects of mining were discussed in your text except:
a. stress on local infrastructure as mining companies attract workers.
b. changes in local land use practices from rural to industrial and urban uses.
c. modification of water quality and air quality regulations to accommodate mine development.
d. economic "boom to bust" as mines become depleted and mining activity declines.

11. Which statement about minerals is false?
a. Minerals are inorganic compounds.
b. Minerals can include metals and non-metals.
c. Minerals contain carbon.
d. Minerals have internal structures that do not change.

12. Minerals can be used in all of the following except:
a. commodity exchange.
b. building/construction.
c. fuel source.
d. life processes.

13. Minerals and metals are concentrated in the earth by all of these processes except:
a. sedimentary processes.
b. parthenogenic processes.
c. igneous processes.
d. biological processes.

14. Dissolved solids (minerals) constitute about ____ of the weight of seawater.
a. 5.0 %
b. 35%
c. 0.53 %
d. 3.5 %
e. 17 %

15. Viable alternatives to compensate for increasingly scarce finite resources include all of the following except:
a. find more resource.
b. recycle what has already been obtained.
c. create more minerals.
d. find a substitute.

Chapter 30 – PLANNING FOR A SUSTAINABLE FUTURE

The Big Picture

Whether or not many of the agricultural, forestry, fishing, industrial, economic and environmental practices we are using today are sustainable cannot be readily determined. Only future generations will know. History has shown many times over that the misuse, misunderstanding, or neglect of the environment that sustains us can lead to ruin. The decline forests on Easter Island and in the ancient Mediterranean region was a major factor leading to the loss of those cultures. By definition, *sustainability* is a concept that entails managing natural resources in a way that maintains the resources for future generations. Clearly, many endeavors undertaken today are unlikely to sustain natural resources and consequently quality of life for future generations. For example, an annual net loss of topsoil, however slight, due to erosion from cropland will result in steadily declining productivity and eventual abandonment of the land. A new world view is needed, nationally and internationally but at local and individual levels as well. In the US and elsewhere, the majority of people place a high value on the environment yet there is often conflict. The perception that environmental protection infringes on individual rights is central to many of these conflicts. In the US, since the mid-1960's environmental laws such as the National Environmental Policy Act have been written to protect the pubic trust- resources that are of value to all. Numerous citizen activist groups have arisen to help safeguard our natural resources by ensuring that environmental laws are effective and enforced. For long-term sustainability, however, government and individuals should take a proactive rather than reactive posture regarding environmental protection. Environmentally sensitive planning is integral to achieving sustainability.

Frequently Asked Questions

To help ensure sustainability, what should be included in a new paradigm or new perspective as to how human interact with the environment?

- The paradigm should include an evolution of new values.
- The paradigm should be inclusive of all peoples.
- The paradigm should be proactive and include planning for change.
- The paradigm should be attractive and based on sound decisions.
- The paradigm should involve environmental justice for all.

Do people really care about sustainability and environmental protection?

- Consistently, polls rank environmental protection as a high priority.

- Polls indicate that many believe environmental protection does not necessarily conflict with economic growth but if it does they are willing to make some economic sacrifices.

Are there examples of sustainable living?
- Yes, but mainly on a small, localized scale.
- Curitiba is a Brazilian community that has seen dramatic improvement in quality of life because its leadership made wise choices for sustainability.
- At the level of the individual, sustainable choices can be made to minimize environmental impacts. For example, the Florida House in Sarasota, Florida illustrates choices in house design and building materials that serves as a model for sustainable living (Figure 30.1).

How is environmental protection addressed in the United States?
- Environmental protection is addressed through federal, state, and local government regulation, international agreements, and through private initiative.
- The United States has a long history of protecting the public trust as well as the rights of private individuals.
- The "**public trust**" consists of those resources of common benefit to all, for which the government is the trustee. An example is the navigable waters of the United States, which cannot be privately owned but are held in public trust.
- The rights of individuals to do as they see fit with their property are established by **common law**. However, if those activities infringe upon rights of adjacent landowners or the public in general, common law may be invoked to protect them. Herein lies one of the most complicated issues regarding environmental protection.

What can be done at the international level to protect the environment?
- Many environmental problems are not contained within the boundaries of nations. Consequently, the international community must address these concerns.
- Overpopulation, global warming, ozone depletion, acid rain and other forms of air pollution, marine pollution, loss of global biodiversity, and the decline of oceanic fisheries are but a few of the global environmental problems that are of international concern.
- International laws are problematic to uphold because sovereign nations cannot enforce laws over other sovereign nations and because nations have different traditions, cultural values, resource bases, and economic status.
- Resolutions to international environmental problems are addressed through international policies, agreements, and treaties.
- The United Nations is a leading forum to address international environmental issues. The U.N. Environmental Summit in 1992 (**Earth Summit**) in Rio de Janeiro was a well-publicized effort to establish international consensus on many environmental issues.
- Two results from the meeting were a concerted attempt to reduce greenhouse gas emissions to 1990 levels by the year 2000 and for developed nations to share environmentally sound technologies with developing nations.

- Many important issues were left unresolved when the Earth Summit concluded, but at a minimum, dialogue between many diverse nations had been established.

What is the role of private citizens and citizens groups in environmental protection?

- Environmental groups (from international and national nonprofit organizations to local citizens action groups) have been very effective in shaping environmental policy and planning.
- One of the most effective means of influencing environmental policy and planning is through non-binding **mediation**. Mediation entails using a neutral facilitator to negotiate a settlement between litigants, and possibly avoiding costly and embittering lawsuits. The mediator helps clarify issues and positions, and gain compromise between the parties involved.
- Efforts of environmental organizations range from active involvement in EIS review, to filing lawsuits, to letter writing campaigns targeting legislative reform. The Sierra Club, The National Wildlife Federation, and The Environmental Defense Fund are examples of mainstream environmental organizations.
- More radical environmental organizations, such as Earth First, Greenpeace, and Sea Shepherd have chosen active protest and organizing boycotts, even *ecotage*, as means to achieve their goals.
- Individual initiative is often overlooked. Individual citizens can play a role in environmental planning and decision making by contacting local, state, and federal representatives, expressing their opinions at public hearings and forums, and using their personal resources to manage privately owned property in an environmentally sound manner.

What federal legislation has been enacted to protect the environmental public trust through environmental impact assessment and planning?

- Key pieces of federal legislation have been enacted in the past three decades (Figure 30.2).
- The **National Environmental Policy Act (NEPA)** specifically addresses environmental impact and planning issues. Under NEPA, an environmental assessment must be prepared for major federal projects that might affect the quality of the human environment. If deemed necessary, a more comprehensive Environmental Impact Statement (**EIS**) may be necessary.
- The goals of NEPA (and similar state legislation) are to focus public and governmental attention on environmental problems that may arise from development projects, provide a framework that will be useful in evaluating the environmental consequences of a project, and generally increase environmental protection efforts nationwide.
- The components of NEPA- EISs are:
  - a summary;
  - a statement of purpose and need;
  - a comparison of alternatives to the project;
  - a description of the environment that will be affected by the project; and
  - a discussion of environmental consequences of the project and alternatives.

- Alternatives that are considered may include **mitigation** actions that are designed to avoid, decrease, or compensate for adverse environmental impacts. For example, in recent years, legal issues pertaining to wetlands have gained much attention. Wetland mitigation is one process used to offset wetland losses due to construction projects. In wetland mitigation, developers may be required to create new wetlands, enhance degraded wetlands, or preserve existing wetlands.
- An important tool now used in environmental impact assessment and planning is **GIS**, (Geographic Information Systems) (Figure 30.5). GIS uses computer modeling to make comparisons and overlays of multiple data sets of environmental variables to assess existing and predicted environmental conditions at a project site.
- EISs can be prepared by governmental agencies or by private environmental consulting companies. In order to maximize objectivity, EISs are subject to review and comment by local, state, and federal agencies, the scientific community, private citizens and citizens groups. Review early in the development of an EIS is called **scoping**.

What role do states have in environmental impact assessment and planning?

- For non-federal projects (state and/or private projects) that do not require federal EISs, many states have legislation (**SEPA**s) which is patterned after NEPA to require an environmental review prior to project initiation.
- The California Environmental Quality Act (CEQA) requires environmental impact reports (EIRs) for many state governmental and private development projects.

What are some examples of the environmental review process?

- An environmental review was conducted at Cape Hatteras National Seashore in North Carolina to determine if the National Park Service should intervene in the natural processes of dune migration and shoreline erosion in order to protect private property and public roads. A management plan was developed that favored allowing natural processes to continue but provided for some limited protection of made-man structures.
- In the highly agricultural San Joaquin valley of central California an environmental review was conducted to assess the consequences of increased selenium concentrations in waters draining from irrigated farmland into the Kesterson Reservoir (part of the Kesterson National Wildlife Refuge. An increased incidence of mortality and deformities in the Refuge's waterfowl populations was attributed to the selenium, a toxic heavy metal that passed through the food chain. Following the review process, the federal Bureau of Reclamation was required to clean up the toxic sites in the Refuge and provide new wetland habitats.

What is land-use planning?

- Land-use planning entails making decisions about how land will be used and implementing those plans to achieve short-term and long-term goals (Figure 30.13).
- Land-use planning proceeds as follows:
  - identify objectives
  - collect, analyze and interpret data

- develop and test alternatives
- formulate land-use plans
- review and adopt plans
- implement plans
- revise and amend plans

- Land use planning has become increasingly important in the U.S. Most land in the U.S. is rural (agricultural, rangeland, forest, or in another natural or quasi-natural state). Only 3% of the land area is urban. Each year, 9000 km^2 of rural land is converted to a new land use category: about half of this area is converted to parks, wilderness areas, wildlife refuges, or recreational area; the remainder is converted to an urban landscape or transportation network (roads and airports).
- In the absence of environmentally sound land-use planning, unrestricted growth tends to lower environmental quality and quality of life for residents.
- Environmentally sound land-use planning is necessary to protect human life and property from natural hazards, such as wildfire, flood, drought, earthquake, and storms (Table 30.2). The effects of many natural hazards are compounded by human activities, for example, flooding frequency and severity in urban areas is often increased as a result of land surfaces being covered with impervious surfaces.
- Effective land-use plans discourage or prohibit inappropriate land use practices in hazard prone areas.

What are different types of land-use plans?

- **Comprehensive planning** is usually accomplished by local planning agencies (city or county) and establishes goals for long-term economic development, public safety, and environmental protection. Comprehensive land-use plans usually contain maps that depict appropriate land uses given such environmental factors as soils type, hydrology, watershed, and natural areas of special concern, as well as infrastructure factors, such as the locations of roads, water and sewer lines, and industrial centers. GIS technology is a highly effective tool in comprehensive land use planning.
- **Urban renewal** involved land-use planning in cities or portions of cities that have undergone decline. A common goal of urban renewal is to preserve the cultural heritage of the area while increasing its livability.
- **Regional planning** entails developing and implementing land-use plans at larger scale, from local and county to multi-county and multi-state. Regional planning often follows watershed boundaries or other natural landscape features. Regional plans typically require coordination by a state or federal agency.
- **Green planning** targets specific global or regional environmental problems. Federal government and international cooperation are usually necessary.
- **Coastal zone planning** was authorized and encouraged in the federal **Coastal Zone Management Act** (1972) and amendments (1976) which gave coastal states the authority to control land use activities within a designated coastal zone. The CZMA was enacted in response to the rapidly deteriorating coastal landscape that was occurring due to uncontrolled development of fragile coastal ecosystems.

How is land-use planning accomplished on public lands?

- Public lands, such as national parks, national forests, and other federal and state lands that are held in public trust require land-use planning.
- Land-use planning on a public land depends largely upon the mission and purpose of the agency managing the site. For example, the mission of the National Park Service is primarily the preservation of the natural and cultural features of the National Parks, whereas the mission of the Forest Service is, in part, maintaining the National Forests for multiple uses.
- Users of National Parks and National Forests have different perceptions as to what services and facilities should be provided and what activities should be allowed. It is difficult to accommodate the interests of all users.

Can environmental impact assessment and land-use planning be accomplished on a global scale?

- Because of the disparate interests, culture, language, economic status, and resources of the various nations, environmental impact assessment and land-use planning on a global scale is very difficult. Establishing dialogue is an important first step.
- The identification of environmental problems and the exploration of possible solutions has been undertaken by international forums such as the United Nations, which sponsored the 1992 "Earth Summit" in Rio de Janeiro, international "think tanks" such as the Club of Rome, international scientific societies, and international conservation organizations.

Ecology In Your Backyard

- Visit your local land-use planning board office. You may be able to obtain a copy of or review a land-use planning document. Planning documents are often made available at public libraries. You should find information about projected road projects, industrial parks, and water and sewer services. Land-use planning offices should also have information regarding zoning ordinances and proposed changes in zoning. Information in planning documents is especially important for prospective home or land buyers. For example, prospective home buyers may wish to re-evaluate their options if the property they are considering is on or adjacent to lands projected for highway construction, water reservoir expansion, industrial complexes, or intensive livestock operations.

Links In The Library

- Kubasek, N. K. and G.S. Silverman. 1994. *Environmental Law*. Prentice-Hall, Englewood Cliffs, NJ.

- Manning, R. 1994. A Good House: Building a Life on the Land. Penguin, New York, NY.
- McHarg, I.L. 1971. *Design With Nature*. Doubleday, Garden City, NY.
- Pilkey, O. and K.L. Dixon. 1996. *The Corps and the Shore*. Island Press, Washington. 256 pp.
- Schulze, P.C. 1996. *Engineering Within Ecological Constraints*. National Academy Press, Washington.

Ecotest

1.Natural resources that are commonly held by all citizens, for which the government is the trustee, are called:
a. the mutual trust.
b. the common trust.
c. the public trust.
d. the inherent trust.

2. If deemed necessary by a preliminary environmental assessment, a comprehensive Environmental Impact Statement (EIS) may be required for federal development projects under what federal legislation?
a. NEPA
b. CERCLA
c. NWPCA
d. NPDES

3. Alternatives to development projects that attempt to avoid, decrease, or compensate for environmental degradation are called ____________ actions.
a. litigation
b. scoping
c. regulatory
d. mitigation

4. The United Nations' role in international environmental impact assessment includes all of the following except:
a. seeking resolutions through treaties.
b. holding international forums to discuss problems and solutions.
c. enforcement actions.
d. sponsoring the 1992 "Earth Summit" in Rio de Janeiro.

5. The use of neutral negotiators to help resolve environmental disputes is called

_________________.

a. mediation

b. ecotage
c. environmentalism
d. mitigation

6. One of several "think tanks" composed of an international group of scientists, former heads of state, and intellectuals is the _____________.
a. Earth First
b. Club of Rome
c. Sierra Club
d. Club Med

7. In the United States, ___ percent of the land is classified as urban.
a. 3
b. 7
c. 12
d. 18

8. _______ land-use plans are typically developed at the city or county level and establish short-term and long-term planning goals.
a. Regional
b. Comprehensive
c. Limited
d. Specific

9. Computer modeling programs that integrate multiple sets of geographic and environmental data thus facilitating land-use decisions are called:
a. Computer Aided Drafting (CAD).
b. Geographic Information Systems (GIS).
c. Geographic Imagery Synthesis (GIS).
d. Geographic Based Design (GBD).

10. The mission statement of the __________ proclaims that lands under their jurisdiction are "mulitple-use" lands.
a. National Park Service
b. National Wildlife Refuge System
c. U.S. Bureau of Land Resources
d. U.S. Forest Service

11. "Common law" refers to all of the following ideas except:
a. the right to do what you see fit with your own property within reason.
b. the right to be protected from landowners whose actions infringe upon your ability to exercise your rights.
c. establishing residence on lands held in the "public trust."
d. All of these are acceptable under common law.

12. EIS stands for:

a. Energy Inspection Survey.
b. Environmental Interment Site.
c. Environmental Impact Statement.
d. Environmental Inspection Service.

13. How much land each year is converted from rural use to another type of use?
a. 90 km^2
b. 100 km^2
c. 9000 km^2
d. 2000 km^2

14. Appropriate land use depends on all of these except:
a. local soil types.
b. hydrology and watershed.
c. infrastructure factors.
d. economic variables, including land cost.

15. What step has been left out before a land-use plan should be formulated? Step one: identify objectives; Step two: collect, analyze and interpret data; Step three: formulate land-use plans.
a. cost of the plan
b. examining alternative strategies
c. implementing the plan
d. planning for unrestricted growth

Ecotest Answers

Chapter 1

1. c
2. d
3. b
4. a
5. a
6. c
7. e
8. b
9. e
10. d
11. b
12. c
13. c
14. d
15. d

Chapter 2

1. e
2. b
3. c
4. b
5. d
6. d
7. a
8. d
9. a
10. b
11. c
12. c
13. c
14. b
15. b

Chapter 3

1. b
2. b
3. a
4. a
5. d
6. d
7. c
8. b
9. d
10. a
11. b
12. d
13. c
14. a
15. d

Chapter 4

1. c
2. d
3. a
4. a
5. a
6. a
7. a
8. a
9. c
10. b
11. d
12. d
13. d
14. e
15. c

Chapter 5

1. a
2. b
3. a
4. b
5. c
6. d
7. c
8. c
9. a
10. c
11. e
12. a
13. d
14. c
15. c

Chapter 6

1. b
2. c
3. a
4. a
5. d
6. b
7. d
8. c
9. a
10. c
11. a
12. b
13. c
14. b
15. a

Chapter 7

1. d
2. e
3. a
4. a
5. b
6. c
7. d
8. a
9. d
10. d
11. b
12. d
13. d
14. d
15. b

Chapter 8

1. b
2. c
3. d
4. a
5. d
6. c
7. c
8. e
9. b
10. d
11. a
12. a
13. b
14. e
15. d

Chapter 9

1. a
2. c
3. d
4. a
5. c
6. a
7. a
8. c
9. d
10. d
11. c
12. c
13. d
14. b
15. d

Chapter 10

1. d
2. c
3. b
4. a
5. b
6. d
7. c
8. b
9. b
10. a
11. a
12. b
13. c
14. d
15. b

Chapter 11

1. c
2. b

3. d
4. a
5. b
6. c
7. d
8. d
9. e
10. d
11. d
12. d
13. c
14. d
15. e

Chapter 12

1. a
2. c
3. b
4. c
5. c
6. c
7. d
8. a
9. b
10. c
11. c
12. c
13. d
14. b
15. b

Chapter 13

1. c
2. c
3. a
4. a
5. a
6. a
7. c
8. d
9. b
10. b
11. c
12. c
13. a
14. b
15. e

Chapter 14

1. a
2. d
3. d
4. d
5. d
6. d
7. d
8. e
9. c
10. c
11. e
12. b
13. c
14. a
15. c

Chapter 15

1. b
2. a
3. b
4. c
5. b
6. a
7. c
8. d
9. c
10. d
11. a
12. b
13. c
14. d
15. a

Chapter 16

1. a
2. a
3. a
4. a
5. b
6. c
7. b
8. e
9. b

10. c
11. b
12. a
13. c
14. a
15. e

Chapter 17

1. c
2. c
3. b
4. a
5. d
6. c
7. d
8. d
9. d
10. a
11. c
12. c
13. a
14. d
15. a

Chapter 18

1. c
2. e
3. b
4. b
5. d
6. d
7. c
8. e
9. a
10. d
11. b
12. d
13. d
14. b
15. a

Chapter 19

1. c
2. b
3. c
4. e
5. c
6. a
7. b
8. d
9. e
10. d
11. b
12. e
13. d
14. b
15. b

Chapter 20

1. c
2. b
3. a
4. b
5. a
6. a
7. e
8. c
9. a
10. a
11. a
12. b
13. b
14. d
15. c

Chapter 21

1 d
2. c
3. b
4. a
5. c
6. b
7. d
8. a
9. a
10. c
11. d
12. b

13. a
14. a
15. a

Chapter 22

1. a
2. b
3. d
4. c
5. c
6. c
7. a
8. a
9. a
10. b
11. c
12. c
13. d
14. d
15. b

Chapter 23

1. c
2. c
3. d
4. d
5. a
6. a
7. b
8. b
9. d
10. a
11. d
12. d
13. c
14. a
15. c

Chapter 24

1. d
2. a
3. d
4. b
5. a
6. b
7. c
8. b
9. a
10. b
11. c
12. b
13. c
14. d
15. e

Chapter 25

1. b
2. c
3. d
4. c
5. a
6. c
7. d
8. a
9. c
10. b
11. c
12. b
13. a
14. d
15. d

Chapter 26

1. b
2. b
3. b
4. b
5. e
6. b
7. b
8. a
9. b
10. c
11. a
12. a
13. a
14. a
15. a

Chapter 27

1. d
2. c
3. b
4. d
5. a
6. c
7. a
8. c
9. a
10. b
11. c
12. d
13. d
14. d
15. d

Chapter 28

1. a
2. d
3. c
4. b
5. a
6. a
7. c
8. a
9. d
10. d
11. a
12. b
13. d
14. b
15. a

Chapter 29

1. d
2. b
3. a
4. d
5. a
6. d
7. c
8. b
9. a
10. c
11. c
12. c
13. b
14. d
15. c

Chapter 30

1. c
2. a
3. d
4. c
5. a
6. b
7. a
8. b
9. b
10. d
11. c
12. c
13. c
14. d
15. b

Notes

Notes

Notes

Notes

Notes

Notes

Notes

Notes

Notes

Notes

Notes

Notes

Notes